▶ Competing against Multinationals in Emerging Markets

Other Palgrave Pivot titles

Nicos Trimikliniotis, Dimitris Parsanoglou and Vassilis S. Tsianos: **Mobile Commons, Migrant Digitalities and the Right to the City**

Claire Westall and Michael Gardiner: **The Public on the Public: The British Public as Trust, Reflexivity and Political Foreclosure**

Federico Caprotti: **Eco-Cities and the Transition to Low Carbon Economies**

Emil Souleimanov and Huseyn Aliyev: **The Individual Disengagement of Avengers, Nationalists, and Jihadists: Why Ex-Militants Choose to Abandon Violence in the North Caucasus**

Scott Austin: **Tao and Trinity: Notes on Self-Reference and the Unity of Opposites in Philosophy**

Shira Chess and Eric Newsom: **Folklore, Horror Stories, and the Slender Man: The Development of an Internet Mythology**

John Hudson, Nam Kyoung Jo and Antonia Keung: **Culture and the Politics of Welfare: Exploring Societal Values and Social Choices**

Paula Loscocco: **Phillis Wheatly's Miltonic Poetics**

Mark Axelrod: **Notions of the Feminine: Literary Essays from Dostoyevsky to Lacan**

John Coyne and Peter Bell: **The Role of Strategic Intelligence in Law Enforcement: Policing Transnational Organized Crime in Canada, the United Kingdom and Australia**

Niall Gildea, Helena Goodwyn, Megan Kitching and Helen Tyson (editors): **English Studies: The State of the Discipline, Past, Present and Future**

Yoel Guzansky: **The Arab Gulf States and Reform in the Middle East: Between Iran and the "Arab Spring"**

Menno Spiering: **A Cultural History of British Euroscepticism**

Matthew Hollow: **Rogue Banking: A History of Financial Fraud in Interwar Britain**

Alexandra Lewis: **Security, Clans and Tribes: Unstable Clans in Somaliland, Yemen and the Gulf of Aden**

Sandy Schumann: **How the Internet Shapes Collective Actions**

Christy M. Oslund: **Disability Services and Disability Studies in Higher Education: History, Contexts, and Social Impacts**

Erika Mansnerus: **Modelling in Public Health Research: How Mathematical Techniques Keep Us Healthy**

William Forbes and Lynn Hodgkinson: **Corporate Governance in the United Kingdom: Past, Present and Future** palgrave▶pivot

Competing against Multinationals in Emerging Markets: Case Studies of SMEs in the Manufacturing Sector

Densil A. Williams

Mona School of Business and Management, University of the West Indies, Jamaica

palgrave
macmillan

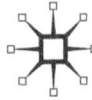

© Densil A. Williams 2015

All rights reserved. No reproduction, copy or transmission of this publication may be made without written permission.

No portion of this publication may be reproduced, copied or transmitted save with written permission or in accordance with the provisions of the Copyright, Designs and Patents Act 1988, or under the terms of any licence permitting limited copying issued by the Copyright Licensing Agency, Saffron House, 6–10 Kirby Street, London EC1N 8TS.

Any person who does any unauthorized act in relation to this publication may be liable to criminal prosecution and civil claims for damages.

The author has asserted his right to be identified as the author of this work in accordance with the Copyright, Designs and Patents Act 1988.

First published 2015 by
PALGRAVE MACMILLAN

Palgrave Macmillan in the UK is an imprint of Macmillan Publishers Limited, registered in England, company number 785998, of Houndmills, Basingstoke, Hampshire RG21 6XS.

Palgrave Macmillan in the US is a division of St Martin's Press LLC, 175 Fifth Avenue, New York, NY 10010.

Palgrave Macmillan is the global academic imprint of the above companies and has companies and representatives throughout the world.

Palgrave® and Macmillan® are registered trademarks in the United States, the United Kingdom, Europe and other countries.

ISBN: 978-1-137-50031-1 EPUB
ISBN: 978-1-137-50032-8 PDF
ISBN: 978-1-137-50030-4 Hardback

A catalogue record for this book is available from the British Library.

A catalog record for this book is available from the Library of Congress.

www.palgrave.com/pivot

DOI: 10.1057/9781137500328

To the hardworking small business owners across the Caribbean and Na'ima Toni-Ann Williams, my daughter

Contents

Preface		vii
Acknowledgements		x
1	Introduction	1
2	Spur Tree Spices	21
3	Yono Industries	35
4	Island Moldings	51
5	Hot Mama's	66
6	Perishables Jamaica Ltd	84
7	Caribbean Flavours and Fragrances	99
8	Conclusions and Lessons Learnt	114
Bibliography		126
Index		130

Preface

The work presented in this volume was motivated by an initiative by *National*,[1] to highlight and publicise the works of committed and diligent new entrepreneurs in the manufacturing sector in Jamaica. The initiative called "The Bold Ones: New Champions of Manufacturing" showcased the work of new entrepreneurs, who are engaged in the production of goods for domestic and export consumptions. National, in highlighting their rationale for this impressive initiative noted that:

> A strong, diverse manufacturing sector provides the best hope for economic recovery and sustained growth and development. We at Continental/National Bakery want to encourage and recognise the brave new entrepreneurs who have plunged into the manufacturing sphere, and in the process, motivate others to join their ranks. National Bakery has sponsored The Bold Ones: New Champions of Manufacturing to encourage small manufacturing companies and to recognise their contribution to the Jamaican economy.

Having recognised these small companies that are engaged in the manufacturing of items that could easily be imported, especially from Asian countries that have greater economies of scale and scope, and can produce similar items far more economically than these companies have done, it raised the question in my mind: Why did these companies decide to engage in manufacturing in the first place? The secondary question then arose: How are they able to survive and prosper when larger and multinational firms can easily produce and sell similar or the same products at much cheaper prices? It is these questions that have motivated the work presented in this book.

Persons may argue that every country will have a few firms that manufacture despite the country's conditions; some items, because of their perishable nature, are vulnerable to international trading. While that may be so, there is still a significant amount of perishable items that are involved in the system of international trade each day. Even packaged meals are now exported and have a shelf life that lasts beyond a few days. This shows that perishability of the items may not be a factor in determining whether or not an item is engaged in international trade.

More importantly, this volume reports on small companies that are domiciled in Caribbean locations (not only Jamaica), that are not perceived to be amenable towards manufacturing, especially where small firms are concerned. In most of these contexts, there exists a high crime rate, which imposes an additional cost of doing business; high interest rates on the cost of capital; unstable exchange rates; and high levels of government bureaucracy as is highlighted by the entrepreneurs in the cases reported in the text. These conditions make it very difficult for business persons to manufacture and sell their outputs competitively to the domestic or even in the international marketplace. The disadvantages of this un-enabling environment is even more pronounced in the case of the small firms, which suffer from the inability to gain economies of scale, and have limited portfolio of products to diversify revenue streams. The owners of these small firms could easily have taken their capital and invested in other areas of the economic value chain, gained even higher returns, and lived a less stressful life. For example, in the case of Jamaica where interest rates on government securities were tendered as high as 52 per cent, small manufacturers could have easily invested their monies in these securities and earned a significant return without having to be bothered about the difficulties of employing persons and running a manufacturing plant. However, despite this attractive alternative, a number of these entrepreneurs decided to turn a blind eye, and have instead tried to build a business that has not only survived but is prospering in the highly competitive manufacturing sector. It is the stories of how these entrepreneurs successfully built these businesses that are reported in this volume. A close reading of the narrative suggests that networking with larger enterprises, brand building, intimate knowledge of the products and line of business, diversification of markets, meeting customer needs, and forward thinking and strategic leadership are at the heart of their survival and prosperity, despite the overwhelming circumstances under which they operate.

The stories presented in this volume are not only inspiring and instructive but they present lessons, which other similar minded persons who want to engage in entrepreneurial pursuits in a tough and competitive industry sector can adopt or adapt in order to enhance the survival and prosperity of their businesses. The lessons are not prescriptive, but describe how other entrepreneurs facing similar conditions in the market place have been able to overcome the various obstacles and build successful enterprises. The narratives presented in the volume give a detailed account of how these entrepreneurs have done it. The use of "first order-narratives" is particularly interesting in an area of research such as this, as it provides insights about the subject under investigation, the persons who are deeply involved, and the entrepreneurs' experiences about how the process works. Narrative as a methodological tool in social enquiry has been gaining significant attention in the academic literature,[2] and so, using it in this form of research on business survival/failure is indeed a novel contribution in and of itself. The narratives described here refer to extended prose, which connects events in a meaningful way and not the strict sociological definition of narratives as a chronological concept.

The book hopes to make a modest contribution to our understanding of the factors that generally drive business success in the small firm. More importantly, it will better help us to understand the drivers of not only survival but prosperity of these small firms with direct evidence from the owners of these enterprises. It will share persons' rich lived experiences and provide insights into the processes and changes overtime that have led to their current state. It is our hope that policymakers who have the responsibility to formulate policies to enhance the survival of small firms, small firm owners who want to build prosperous enterprises, and business practitioners who are engaged in training small firms to improve their chances of success, will find the lessons outlined in these pages useful.

Notes

1 National is a company in the manufacturing sector that produces light food products such as breads, biscuits, buns, and so on. It is owned and operated by the Hendrickson's, one of Jamaica's most entrepreneurial families.
2 See Elliott, J., 2005. *Using Narrative in Social Research: Qualitative and Quantitative Approaches*, London: Sage, pp. 1–199, for a good discussion on the developments in the use of narrative in social research.

Acknowledgements

The research and writing that go into the production of a book is not an effortless task. Given the size of the project and the numerous tasks involved, it requires tremendous contribution of time and energy from a wide cross section of stakeholders.

I would like to thank the many small business owners who have given of their time to be interviewed for this volume. They are extremely busy individuals but they have dedicated the time to sit for hours to be interviewed. This volume would not be possible without those efforts.

My research assistant, Mr Edward Dixon, has been exceptional in helping to put this volume together. His enthusiasm and strong work ethic were valuable in getting this project to fruition. I thank him immensely.

Professor, the Hon. Gordon Shirley former Principal of the UWI, Mona Campus was instrumental in helping me to shape the ideas for this book. We exchanged ideas on the form and content of the material in the conceptualisation phase of the work. Those exchanges proved invaluable to the final output. I thank him, immensely.

Financial resources are critical in getting a project like this off the ground. I thank the office of the principal at UWI, Mona for the New Initiative Grant which enabled me to fund the research for this book. Those resources were valuable in the research phase of the work for this volume.

I thank my friends from both sides of the Atlantic who have always been interested in my work and have listened to my many iterations of the work to bring it to this level. I thank you for your patience. I will also like to thank the anonymous referees for the helpful comments which have improved the quality of the work.

Finally, I would like to thank Tracy and Na'ima for their patience and understanding while I spent inordinately long periods away from home in preparation of this volume. I thank my mother, father, and siblings for their continued moral support.

Despite the many persons who have contributed to this volume, I take full responsibility for any lacunae.

palgrave▸**pivot**

www.palgrave.com/pivot

1
Introduction

Abstract: *This chapter details an account of the theoretical lenses which are used to guide the development of the arguments in this book. It focuses on efficiency theory, limited portfolio theory and market structure theory in addition to the resource-based view of the firm to explain how small firms can compete against multinational firms in the manufacturing sector in emerging economies and survive. Among other things, the chapter also provides an overview of the manufacturing sector and shows how the cases are developed using the narrative methodology. It provides a good description of narrative as a method that can be used effectively in business research.*

Williams, Densil A. *Competing against Multinationals in Emerging Markets: Case Studies of SMEs in the Manufacturing Sector.* Basingstoke: Palgrave Macmillan, 2015. DOI: 10.1057/9781137500328.0004.

Competing against Multinationals in Emerging Markets: Case Studies of SMEs in the Manufacturing Sector is a collection of case studies of small firms which are independently owned and managed, that have overcome obstacles the highly competitive industry/sectors in which they operate and remain open after initial start-up.[1] The entrepreneurship literature is replete with evidence of small firms starting and operating for only a few years and then disappearing from the market place (GEM 2012). This is even more so for those firms that operate in highly competitive sectors (such as manufacturing) that require scale and scope for survival. The cases presented in this volume provide a comprehensive look at the origins and expansion of small firms to the future of the business among other areas such as customers, their operation in the business environment, and also, their processes and markets. The idea is to provide a first-hand account of how the entrepreneurs have managed and operated their enterprises amidst the turbulent environment in which they operate, and still are able to survive and grow their businesses. The common lessons from these experiences will provide significant insights to both policymakers who are interested in moving the small firms from merely start-ups to established operations, and also to other business owners who need to identify the optimum mix of strategies to move their firms from start-ups to established operations. The lessons are inspirational as well as insightful.

This focus on the small firm is compelling because there is a general bias in the extant literature that manufacturing is the domain of large and well-established firms (Page 2009). It appears that small firms have no place in this area of the economic value chain. Further, academics who pursue this line of work, are normally looked at askance; because, as the argument goes, the lessons from larger firms can be used to inform policy decisions of smaller firms. Hall (1995) pointed out in a very clear and candid way why this thinking is not only wrong but also dangerous. The small firm is not a scaled down version of large firms. In fact, as Hall pointed out, it may even be more useful to study small firms, which are much simpler organisational forms that could provide insights into issues that affect larger firms, which are more complex. This is similar to biological research, where the study of simple organisms helps scientists to draw conclusions about more complex ones. Similarly, the impact of the small firm on the national economies, in terms of employment and output, have given rise to an increased interest from policymakers on issues that affect small firms. It would therefore be unwise for academic disciplines, especially in the field of management and business studies, not to pay particular attention to the issues that have implications for the growth and expansion of these firms. The works covered in this book will make a modest contribution to the understanding of this vital area of survival and growth of small firms.

This chapter among other things will give some insights into the general academic literature on the survival and failure of the firm, provide a general overview of the global manufacturing sector, and locate the Caribbean realities within the wider global context. It will also provide an overview as to the approach that was used to derive the cases.

Survival and failure in the context of entrepreneurship research

While there is unlikely to be a single explanation for the survival or failure of small firms, scholars who are keen on understanding the survival/failure phenomenon from the perspective of the smaller firms, have offered various explanations. One of the major findings in most empirical works on survival and failure suggests that firm size and age are strong explanatory factors. Indeed, early studies on the phenomenon suggest that there is a positive correlation between firm size, age, and survival. The converse is also true, that is, there is a negative relationship between size, age, and firm failure.[2] While empirically these results have been mixed, theoretically, the explanations are rich.

To explain why size and age are critical to firm survival or failure, Jovanovic (1982) advanced an explanation using the efficiency argument. Jovanovic argued that firms will make output decisions based on their levels of efficiency. As such, if a firm learns to become more efficient over time, it will lead to a reduction in the cost of production, and as such, the firm will produce more outputs. All other things being equal, these firms will survive in the market place. Similarly, firms that have not become more efficient will face a higher cost structure, and as such will cut back on output. Once the cost goes beyond a certain level, the firm will be forced to close its operations. The critical insight from Jovanovic therefore, is that, as the firm grows in size, it will be able to learn to become more efficient and through the learning curve effect, will be able to reduce cost and remain in business.

While the efficiency argument had some merit, other scholars have also proffered different rationale for the survival/failure of small firms. Using insights from economics on economies of scale and scope, Hall (1995) argued that high failure rates among small firms generally result from the low level of opportunities available to these firms mainly because of what he called, *limited portfolio*. Small firms he argued are limited in terms of the products they offer, and also the markets in which they operate. Because of these limitations, the small firm is unable to benefit from economies of scale, which could bring greater efficiency and reduce cost; and also, economies of scope, which would diversify their cost across a wider range of products and markets thus reducing the risk of over dependence on a singular product or market. It would suggest therefore that small firms that are not able to benefit from these advantages will see a higher level of failure than larger firms, which have a more expansive portfolio and can generate scale and scope, thus surviving longer.

Market structure is also another explanation put forward as to why small firms generally have a higher failure rate than larger firms. Simply put, the markets and more precisely, the market segments in which small firms operate generally have low barriers to entry. If these market segments are deemed lucrative, then others can easily set up operations as well (Hall 1995). The market segment can easily become overcrowded and as such, the level of demand that each firm receives also reduces. This may lead to cost of operations outstripping demand with the firm ending up in financial difficulties and having to close its doors. So, once the barriers to entry and exit are not particularly high, the smaller firms may not have much option when markets are highly concentrated with a large number of players, than to exit the segment if their demand is too low to meet the requirements of their cost structure.

Resources and survival/failure

A significant body of work exists on the factors that lead to survival (and by inference, failure) of small firms. Several researchers used the resource-based view of the firm as the theoretical lens through which to understand the research problem (Williams 2014; Ahmad & Seet 2009; Semrau & Werner 2012; Watson 2007; Campbell et al. 2012). Similar to the limited portfolio argument espoused by Hall (1995), this body of work argues that small firms that normally fail, have a limited amount of resources, and therefore, cannot enjoy the benefits of economies of scale and gain lower cost. So the corollary is also true; the argument suggests that when these firms have a large stock of resources, they are more than likely to benefit from economies of scale, which will lead to lower cost of doing business, and so, they can produce more output. This resource-based argument is also similar to Jovanovic's (1982) work, which showed that size and survival of the firm are positively correlated. Theoretically, the literature is almost at one in terms of explaining survival/failure of the firm using the resource-based argument as first espoused by Penrose (1959) and further articulated by Barney (1991).

Despite the near theoretical consensus on the role of resources in explaining business survival/failure, the empirical results tend to have different outcomes. When modelling the impact of resources on the failure or survival of a firm, in some cases, the model shows different relationships among variables. For example, some models reveal that there is a positive relationship between age and survival (using profitability as a measure) of a firm, while similar studies have shown a negative relationship (Watson 2007). A closer reading of the works in this area however, revealed that the way in which different authors measure resources can have an impact on the outcome that is derived from various studies. In spite of the controversies surrounding the relationship of surrogates of resources and business survival, there are still a number of proxy variables that are used to capture resources, for which there appears to be general consensus on the relationship with survival.

One of the critical proxy variables that is used to capture resources in the firm, and for which there is almost general consensus that it enhances business survival, is *networks* (Watson 2007; Semrau & Werner 2012; Lee et al. 2012; Aldrich & Reese 1993). A network, which is generally defined as inter-firm relationships (firm to firm or even in the owners own social connections), is quite a useful asset that small firms can leverage to overcome their limited portfolio of resources. The network can be used to help the small firm to overcome resource limitations; as it can allow the firms to gain valuable and necessary resources such as market knowledge, financial support, or human resources support to build its competitive advantage (Liao & Welsh 2005; Kiss et al. 2012). In addition, other researchers have also noted that networks can provide critical information, which is important for decision making but is costly to obtain (Coleman 1988). Attaining

DOI: 10.1057/9781137500328.0004

these resources through a firm's network relationship can prove to be a considerable benefit to small firms, which under normal circumstances would not have the financial capital to acquire them in the first place. Gaining proprietary information from one's network can help to make the difference between firms that survive and those that fail (Lee et al. 2012; Watson 2007). Indeed, the size of the network can greatly impact on the level of competitive advantage, which the firm will gain from this interaction. For example, firms that have very small networks with only a few friends and maybe family members will find that their access to information may not be as rich as those firms and owners with a wider network of friends, family members, business associates, and advisors (Watson 2007; Granovetter 1983).

While the literature in general supports the notion of a positive role for networks in enhancing business survival, there is still caution as to the extent to which firms should seek out networks. In an insightful study carried out by Semrau and Werner (2012), they argued that while networks provide significant benefits for venture creation, there is an opportunity cost that comes with getting involved in networks. Importantly, in trying to build a network, owners of firms have to invest a significant amount of time to both establish the network and also maintain it. Therefore, if the network is not properly conceptualised, the payoff of this investment may not always be positive for the firm, as the time could be used to do other higher value activities. It is for this reason, Semrau and Werner (2012) argued, that the relationship between networks and firm performance, specifically the founding of new ventures may not be a linear one but an inverted U-shape. The interpretation of this U-shaped relationship is that at some point, the time and energy invested in building and maintaining network relationships leads to a negative rate on investment. Therefore, it is not always true that extensive networks and intense network relationships will generally lead to strong firm performance (Davidsson & Honig 2003). This is indeed an important observation to note as most persons will also think that all collaborations are good for businesses, especially small businesses.

Other researchers have also found similar results that networks are not always good for businesses if they are not constructed well (Watson 2007). In a very thorough study of the role of networks in helping businesses to survive, Watson (2007) highlighted the role of networks in determining firm performance for established enterprises. The findings suggest that there might be some optimum level of resources that an owner should devote to networking. In trying to quantify the optimal level, Watson (2007) noted that accessing more than six networks during a year is likely to be counter-productive. Additionally, he went on to look at the networks of the owners of these firms as well, and suggested that accessing any individual network on more than three occasions during a year is also likely to be counter-productive. The results from his detailed investigation show that both formal and informal networks are associated with firm survival. Further, as it relates to individual networks, it was found that accessing advice from external accountants was the only network source positively associated with firm survival and growth. Indeed, similar to other researchers, Watson (2007) found that network intensity was more critical to firm survival than network range, and the opposite was true for firm growth. In line with the discourse on firm age and survival of the firm as espoused in Hall (1995), Watson also found that networking was equally important to both young and old firms. Overall, similar to Semrau and Werner (2012), Watson (2007) argued that given that business failure generally results in heavy personal loss, owners need to seriously consider the range and intensity with which they access various potential networks, whether formal or informal.

Similar to Watson's (2007) detailed work on the role of networks and their impact on firm performance, a number of other empirical works have also highlighted the relationship between firm performance and networks, with performance generally being measured as survival/failure. Professional advice as a proxy for networks was found to have a positive relationship with performance (Duchesneau & Gartner 1990). Also, other works have argued that successful companies relied on accountants' information and advice compared to unsuccessful companies (Watson 2007). Some studies have also found that the financial performance of firms is positively related to using external management advisory services. Researchers have also shown that network development, particularly at the national and international levels, was positively associated with firm growth (Donckels & Lambrecht 1995). In addition, Lerner et al. (1997) argued that network affiliation was significantly related to profitability, and that the use of outside advisors was related to revenue. Also, Larsson et al. (2003) found that a lack of contacts with outside expert advisors hindered small business expansion. In their very insightful work, Carter et al. (1996) found that the wider the networks consultations, especially professional advisors, the more likely the owner of the firm is to succeed in securing equity financing.

The relationship between networks and firm performance is not always positive however. A number of studies have identified deviations from the conventional wisdom that networks are always positive for firm survival. Aldrich and Reese (1993) and Cooper et al. (1994) did not find a positive relationship between networking and firm performance. Again, their work showed that measurement of the two important constructs (networks and performance) are critical to the outcome of any research in this area of work.

Despite the opposite findings by a handful of authors on the role of networks and firm performance especially when performance is measured as survival/failure, the majority of the arguments seem to be more compelling in the direction of a positive relationship between networks and business success, in the form of external advice from accountants, the embedded relationships that the firm has with other firms, the informal ties that the owner has with professional groups, and a positive relationship with the performance of the firm. The interpretation therefore, is that, all other things being equal, small firms that have fewer networks and network contacts should see a higher failure rate than those which get advice from external sources and have stronger network contacts. The rationale seems to be that they will have fewer resources with which to work and to facilitate economies of scale in production and distribution. Accordingly, they will suffer from a higher cost structure, and the risk of doing business will be much higher than firms with larger amounts of resources that can generate economies of scale in production and distribution; thereby lowering their operating costs, becoming more efficient and producing a larger amount of outputs. And, with all other things being equal, growing the business.

Similar to networks, which is a surrogate for the theoretical foundation of the resource-based view of the firm, the efficiency arguments put forward by Jovanovic, the limited portfolio and market structure arguments advanced by Hall, other important variables have also been highlighted to explain firm survival/failure. It is beyond the scope of this work to give a comprehensive review of all those variables. However, in keeping within the discipline boundaries of the work (economics, strategic management, and international business), the work presented in this volume has tried to analyse the most common variables that impact on survival/failure of the firm as viewed through the theoretical lenses that underpin this work. In this regard, other important variables such as owners' perceptions of failure, size, and age, and their impact on failure/survival of the firm were also examined in the wider review of the literature on business survival/failure. For a detail review of the results from these works readers can consult Hall (1995), Watson (2007), and Williams (2014) among other sources.

Summarily, one can argue that similar to the discourse on economic growth of a country, there is no single variable that can explain why firms survive or fail. The domains of sociology and psychology looking at owners' perceptions of survival and failure can provide insights into the issue. For example, economics through its focus on economies of scale, market structure, and efficiency among other areas can provide a view; strategic management through its focus on the resource-based view of the firm provides enormous insights; international business through its focus on portfolio management has also provided strong insights. Equally, the emerging discipline of entrepreneurship through its focus on the owner of the firm, the entrepreneur, has provided strong insights on business survival/failure. The works presented in this volume have drawn on all these various disciplines to shed light on the survival of small firms in the manufacturing sector, a sector in which they have all expected to fail; especially given their liability of size and inability to generate scale, which is crucial for survival in that sector.

Overview of the manufacturing sector

The review will seek to examine some of the global trends and major topical issues as it relates to the manufacturing sector. In particular, issues such as: the role of manufacturing in international trade; manufacturing and the services sector; labour-related issues in the sector; manufacturing and Small and Medium Enterprises (SMEs); environmental issues; current practices in responses to changes in international trade, and the future challenges to the growth of the sector.

Definitional issues

The importance of the role and function of the manufacturing sector to an economy has often been highly contested (Tregenna 2011; Sharp et al. 1999; Page 2009). Commentators generally refer to the sector as the "engine of growth" for an economy, and there seems to be some general consensus around this notion. This description of the sector and its importance to economic growth has had a long history dating back to the Industrial Revolution of the 18th century (Sharp et al. 1999). Scholars have argued that a conceptual distinction must be made between manufacturing and industry, as both terms are often used interchangeably without due regard to their differences (Tregenna 2011). Tregenna (2011) argued that industry is a more comprehensive concept, which also encapsulates mining and construction. On the other hand, manufacturing has a series of specific characteristics, which have given rise to it being described as the "engine of growth" for a nation. Citing empirical research on Kaldor's Law, Tregenna supported the argument of the role played by manufacturing as an engine of growth, by arguing that the sector is special because of its dynamic economies of scales, backward and forward linkages with other sectors in the economy, strong experiential learning component, and high levels of innovation and progress among other factors (Tregenna 2011). The works presented in this book will not try to settle this controversy as to the definition of manufacturing versus industry. That is beyond the scope of this volume. The volume will continue to use manufacturing in the context for which it is intended based on the description given in the trade accounts of the countries studied in this volume.

International trade and the manufacturing sector

The impact of the international trade regime has been noticeable on the manufacturing sector, with many theorists citing the 1970s and 1980s as being one of the watershed periods (Jenkins & Sen 2006). Jenkins and Sen argued that the 1980s saw a shift away from predominantly North-North trade to it being increasingly North-South, as many developing countries emerged as major exporters of manufactured products in the aftermath of the decolonisation period. Further, Page (2009) argued that this expansion has continued in recent times with developing countries increasing their share of the global manufacturing by almost 7 per cent between 2000 and 2005. This is in comparison to a 1.1 per cent growth in developed economies in the corresponding period. An example of this growth has been in South Africa where the manufacturing sector has become the largest contributor to the country's gross domestic product (GDP) through its potential to generate employment as well as to enhance the country's economic growth (Davies 2001).

In addition to having a shift in the level of growth among developing countries, there have been some changes at the local level resulting from trade liberalisation, especially since the 1980s, leading to greater levels of productivity of domestic firms in both the formal and the informal economic space (Nataraj 2011). Nataraj argued that whilst the small informal firms are unable to compete with the international as well as larger, high-end firms in India, liberalisation has ensured that they have been able to secure a larger section of the domestic market. This is also true for other countries that are comparatively smaller than the BRICS[3] (Terziovski 2010). Indeed, Terziovski (2010) suggested that further research needs to be done in order for a greater appreciation to be had of the nature of the operations of these firms as well as to provide a better platform for the development and support of these small firms in order for them to benefit from the purported "gains" from international trade.

Similarly, Tybout (2000) argued that these policies implemented by governments in developing countries normally favour large firms, which are inhibiting factors to the development of small firms. He pointed out that investment incentives are available for projects above a minimum scale and large scale producers are singled out for special subsidies (Tybout 2000). Others argue that it is the absence of financial support as well as institutional underdevelopment that are to be blamed for the current status of SMEs when compared to that of the larger firms in the sector (Ayyagari et al. 2007).

SMEs in the manufacturing sector

There is a lack of in-depth research and acknowledgement of the contribution made by SMEs to the manufacturing sector specifically, and to international trade generally (Ayyagari et al. 2007). In their research of the contribution made by the SMEs to the manufacturing sector and the GDP in 76 countries, Ayyagari et al. (2007) argued that the development of a robust SME sector in manufacturing was associated with a business environment which facilitates entry, establishes property rights, and fosters access to external finances. They also found that there was a positive relationship between the share of SMEs in manufacturing and the GDP per capita growth. They further argued that the contribution of the SMEs could also be viewed through the lens of employment creation.

Manufacturing sector and its impact on the economy

The discourse on the impact of the manufacturing sector on the economy of a country, particularly employment, has been one of the main points of discussion for economists and public policymakers for a long time (Pilat et al. 2006; Jenkins & Sen 2006; Davies 2001). It has been argued by some theorists that the manufacturing sector has the greatest potential to generate employment opportunities and thereby enhance economic growth (Davies 2001). Others have been more measured in their analysis of the current role played by the manufacturing sector in particular countries. They have pointed out that the interaction between the country and its relative position in the global economy are important contextual issues that must be examined before any definitive statement can be made (Jenkins & Sen 2006).

In their study, Pilat et al. (2006) offered support to the view of Jenkins and Sen by arguing that in recent years, OECD countries have generally registered declines in the share of manufacturing in overall employment when compared to other non-OECD countries.

They further argued that the reason for the decline in the number of manufacturing workers in virtually in all OECD countries, with the noticeable exception of Canada, Ireland, Mexico, New Zealand, and Spain, is the rapid employment growth in the services sector.

Scholars have argued that there are a number of factors which should be critically examined in this regard. Firstly, there has been an uneven distribution of the gains in manufacturing among the OECD countries with China and Russia being impacted by the economic restructuring, which has seen the gradual elimination of the inefficient state-owned plants; and Latin America's and Africa's growth has either been on the decline or remain at low levels (Pilat et al. 2006). This, according to Page (2009) especially in Africa, is due to the lack of diversity, the current export structure of the sector as well as the decreasing sophistication in the manufacturing sector in the country. Further, other scholars have argued that the reduction in the manufacturing sector's contribution to employment results from the fact that the character of the work in the manufacturing sector in OECD countries has also changed, moving further up the value chain and away from the production of purely primary products. This might have also resulted in workers being reclassified in the service sector as opposed to the manufacturing sector (Jenkins & Sen 2006; Pilat et al. 2006).

Additionally, other scholars have attempted to analyse the contribution of manufacturing to the economy from the perspective of the SMEs (both formal and informal) in over 76 countries, and argued that the dimension of the business environment was critical for the survival of

Introduction 13

SMEs in this sector especially the informal SMEs. The conclusion is not dissimilar for developed countries with more mature manufacturing sectors (Houseman et al. 2011).

The change in wages is also an explanation put forward for the declining contribution of the manufacturing sector to a country's GDP. Writers have argued that there is a change in the relative wages of workers in the sector as compared with workers in other sectors such as the services sector thus causing persons to switch sectors (Pilat et al. 2006). Other researchers have also paralleled the decline in the manufacturing industry with the rise of the services sector in the early 1970s (Sharp et al. 1999). Some have posited that the reason has been due to the fact that international capital flows have shifted away from traditional manufacturing towards services (Doytch & Uctum 2011; Baldwin & Evenett 2012). Others have attributed the change partially to the impact of the Information and Communications Technology (ICTs) on both manufacturing and services (Sharp et al. 1999).

Manufacturing and the services sector

What is clear from the discourse is that there has been a shift in most economies from manufacturing to services. It has been recorded that the contribution of services to the economy, as represented in the GDP, is now at 70 per cent (Doytch & Uctum 2011). Despite the reasons proffered for the switch, experts have observed that the rise in the size of the services sector, especially in developed economies, has resulted in the advancements in productivity and technology (Kozicki 1997). Other authors have claimed that there has been a "servitisation" of the manufacturing sector (Baines et al. 2009). The current trend in manufacturing has been a move toward providing a mix of services aligned to the manufactured product. This includes a "combination of goods, services, support, self-service and knowledge in order to add value to core product offerings" (Baines et al. 2009, p. 554). Others have added that four factors tend to account for the chasm that has developed between manufacturing and services. These include: outsourcing, measurement issues, lagging computerisation, and differences in competitive pressures (Kozicki 1997).

Significant changes and developments in the manufacturing sector

The myriad of changes in the international trading environment that have occurred both at the global level and the concomitant effects at the local level have been observed by many scholars as massive (Baldwin & Evenett 2012). It is important to highlight that some of these have been as a result of globalisation whilst others have not. However, the impact on areas such as technology, and on organisations is noteworthy (Baldwin & Evenett 2012). Policymakers at the national level as well as those at the firm level have designed systems and procedures to accommodate these changes. How organisations respond will vary according to whether the impact is felt across the industry (industry effect) or within the industry (firm effect) (Thomas & D'Aveni 2009). Davies has argued in favour of four major strategies that have been employed in an effort to improve manufacturing competitiveness. These include: industrial specialisation and movement up the value chain, benefaction of natural resources, targeting of their capabilities, and more recently, clustering.

Some of the changes that have occurred have been as a result of the lowering of barriers to trade (Baldwin & Evenett 2012). This has resulted in greater levels of competition to be increased for both large and small firms within different countries (Karim et al. 2008). Others cite high turnover rates among firms due to exogenous shocks, such as regulatory changes engineered from the level of global trade that impact on productivity (Newman et al. 2011). Still, others have cited studies, which indicate that the competitive landscape is less stable and firm advantages are less lasting (Thomas & D'Aveni 2009). These challenges have resulted in firms, both large and small, adopting strategies geared at combating the effects. At the international level, there has been a development in the concept of "vertical specialisation", whereby there has been an increasing interconnectedness of production processes, which stretch across many countries with each specialising in a particular stage of a goods production sequence (Hummels et al. 2001). Other firms have resorted to outsourcing as well as other technological or organisational changes in order to remain competitive (Baldwin & Evenett 2012). Further, others have resorted to "industry switching" (Newman et al. 2011). Similarly, others have had to implement product and process innovation as well as focusing on producing quality, new products and so on (Karim et al. 2008). In some countries, there have been more inward looking approaches with countries like the US adopting a policy of "offshoring", which essentially is the substitution of imported fair domestically produced goods and services (Houseman et al. 2011).

SMEs have also been affected and have thus had to make adjustments in their operations in order to remain viable. The automotive sector is an example of this trend whereby major automobile component suppliers have had to set up manufacturing facilities where their customers are also based (Lee et al. 2012). These new developments have resulted in a significant number of mergers, take-overs, and alliances as well as the "sub-contracting" of non-core activities down the supply chain by the vehicle manufacturers and has led to a concentration of supplier companies into a number of tiers (Lee et al. 2012).

For SMEs to cope with the rapid changes in the sector, scholars have argued that there needs to be a cost-based strategy adopted by SMEs to combat competitively priced products from countries such as China and India (Terziovski 2010). This strategy, it is argued, is easier to adopt for SMEs, where the chain of command ends with the CEO, than for larger companies, where the decision-making process is more complex.

The cost of energy, which is a debate that is perennial to most manufacturers in the Caribbean region has also attracted attention in the wider literature on manufacturing (Ramirez et al. 2005; Bassi et al. 2012). Indeed, Ramirez et al. (2005) contended that the issue of energy has had a long history in the manufacturing sector, with energy efficiency being an important subject for discussion at the political and technological levels. They maintained that the shocks to international trade, and the concomitant effect of the oil crises were important signposts in the manufacturing sector, and to which actors in the industry are constantly reminded. Further, there is a constant reminder of the potential impact of these and other exogenous shocks on the firms and the countries. Additionally, the discussions surrounding the reduction of energy-related carbon emissions have become increasingly important to the discussions in light of the Kyoto Protocol on greenhouse gas emission. This demonstrates the level of economic and environmental dependence of the global manufacturing industry on fossil fuels (Bassi et al. 2012). It is apparent that conversations going forward must now begin to factor in the move away from this dependence to more environmentally friendly and sustainable methods of production, which will wean the sector from its overreliance on volatile fossil fuel, while at the same time allaying the fears of a vocal environmental lobby group.

In summary, the manufacturing sector has seen significant changes over the years, and firms that operate in the sector and its various sub-sectors have had to make strategic adjustments in order to survive. The impacts, as the literature showed, were not only at the level of large firms but small and micro firms that operate in the sector as well. The cases presented in this volume will show how a set of smaller firms, which have had to grapple with the changes highlighted, were able to strategise and implement plans in order to survive the ebb and flow of the sector. The lessons to be learnt from these cases will be useful to policymakers who are interested in carving out policies that will help more SMEs to survive in highly competitive sectors like manufacturing. Further, policymakers and managers in these small firms will be able to extrapolate lessons that can guide their own operations and make their firms more successful.

The development of the cases

A very structured process was used to develop the cases presented in this volume. To ensure the work meets a high standard of academic rigour and provides practical guidance to the small business community of practitioners and policymakers, the case development followed the guidelines as set out in Yin (2003, p. 86) as closely as possible. While efforts were made to get a representative sample of cases, given the willingness of firms to participate in the study, we resorted to a convenience sample of those firms that were willing and met the criteria for study.

The companies selected from this sub-sector for analysis had to meet the following criteria:

1. Have a workforce of no more than 250 employees
2. Must be indigenously owned
3. Must have been in existence for more than 10 years
4. Should at least be exporting and to more than one market
5. Should have some sales outside of the Caribbean region

These criteria are aimed at ensuring that the firm is not only being protected by local policies but that they have the necessary expertise to compete overseas and win market share as well. This is vital to ensure their long-term survival as the manufacturing sector becomes more competitive through either importation of foreign products sold locally or the competition in export markets for the products.

To facilitate the process of writing the cases for this volume, four non-mutually exclusive steps were followed. These include: documentary review, data collection, data analyses, and the writing of the cases.

Documentary review

This stage involved the researchers engaging in a review of both the academic literature on survival and failure of SMEs and a review of company documents for the firms that agreed to participate in the study. This phase helped to provide historical background on the companies, background information on the sub-sector in which they operate, and also other information that might be relevant to the case. It must be noted that in most cases, these small firms do not have documented evidence on their operations so information was very sparse in this regard. Some companies did not even have a functional website, which showed historical information about the companies. In some cases, evidence was taken from archival sources such as the Jamaica Gleaner and other print media.

Data collection

The data collection phase of the project involved comprehensive, open-ended interviews with owners (entrepreneurs), managers, employees, and other principals of the firms. In all cases, the interviews were tape recorded and also photographs taken of the operations of the company. Most interviews lasted for a minimum of 90 minutes and in some cases, interviews exceeded 120 minutes. This is in line with the recommendations put forward by many scholars who have been using narrative as a methodological concept for understanding persons lived experiences (Elliott 2005).

Interviews were done with the entrepreneur (owner) first, then with staff and other principals of the firm. In some cases, close to the end of the interviewing process, the entire staff complement also joined the interviews along with their managers to give their impression of the company and to take pictures of the operations. Besides interviews for primary data collection, secondary data from company documents were perused to gain additional information based on points raised in the interviews. The interviewers were also able to observe the operations of the firm while the interview was taking place. This allowed for ambiguous statements to be clarified.

Data analyses

The data generated from the interviews and secondary sources were analysed by the researchers using standard techniques in the field of qualitative research. Specifically, the data gathering and analysis followed closely, the narrative methodological framework (Elliot 2005). Narrative as used in this context refers to the extended prose, which outlines the discourse with the interviewee in a sequential order, and in a meaningful way for the audience which this book hopes to reach. The research presented in this volume was more interested in understanding the lived experiences of the owners of the small firms and how this experience has helped the firm to survive. This sort of discovery process falls squarely within the conceptual nature of narrative analysis (Elliot 2005). Consequently, the meaning of narrative as restricted to the temporal definition of chronology, which heavily reflects the sociological operationalisation of the construct will not be used in this text. Indeed, the text will present what Carr (1997) referred to as "first order narratives", which is defined as stories that individuals tell about themselves and their own experiences.

The recorded interviews were transcribed into large volumes of data and then checked for accuracy especially with dates and figures.[4] The research then analysed the volume of records to identify thematic groups under which certain responses may fall. This led to a number of thematic areas as evidenced in the written cases. The direct narrative evidenced from the cases was also used to buttress the researcher's interpretations of the findings, and to provide a sense of the direct involvement of the entrepreneurs in the cases. To ensure accuracy and proper representation of the information, the researcher, after developing the cases, sent them back to the principals of the firms to ensure that the data are reflected accurately and that the essence of their company is truly represented. Where there were discrepancies, the recorded information was referred to for clarification.

Development and writing of the case

The information gathered from the analyses of the data was used to inform the writing of the case material. The cases will present material for discussion purposes that will help managers in SMEs and public policymakers to think more strategically about the things that they will need to do in order to ensure the survival of their firms in a highly competitive environment. The cases are not so much organised to provide insights into a specific problem that the organisation may be facing. From the narratives presented, specific areas of weaknesses will become apparent and discussants can then determine how best to apply various analytical tools to solve these problems. The lessons learnt from these cases, it is hoped, will help SME owners and managers to avoid past mistakes when they are considering the future expansion of their firms or the development of new business opportunities. It should be noted that the cases in this volume will not serve as a prescription for how other small businesses should manage their operations in order to survive highly competitive environments. Instead, the lessons learnt can be used as a guide to developing "best practices" for managing in highly competitive environments. For as any entrepreneur will quickly remind you, all firms will have their idiosyncratic features that will require novel approaches to management. Survival and prosperity involve both an art and a science.

Organisation of the text

Following this introductory chapter, the remaining chapters of the text generally represent cases that are drawn from the various companies. These cases look at various aspects of the firm's operations from its origins and expansion, its customers and employees, its products and production processes to its future as contemplated by the owner/entrepreneur. Each chapter begins with an abstract which outlines the content of the chapter and makes it easier for readers to get a view of the components contained in the chapter. The book ends with a concluding chapter, which focuses on the lessons that can be drawn from the cases and some recommendations that SME owners can look into for the improvement of their own operations. The recommendations are not meant to be exhaustive but reflect the interviewer's own observations of possible areas of weakness that the firms may want to consider improving on.

Concluding remarks

This chapter provided an overview of the academic literature on survival and failure of the firm drawing on various theoretical and empirical explanations to explain the phenomenon. Efficiency theory, limited portfolio, and market structure explanations were considered to be useful lenses through which to view survival/failure of the small firm.

In addition, the resource-based view of the firm was also considered. Further, the chapter highlighted the various developments in the manufacturing sector and showed the implications of these developments on the operations of the small firm. Subsequent to this, the chapter then showed the method behind the development and writing of the cases. The cases presented in the next section will reflect the outputs from the interviewing process.

Notes

1. We refer to independently owned in order to ensure that the there is no bias in the survival of small firms depending on the benevolence of a larger and more resource rich enterprise. As such, this study was limited to firms that are owner-managed (the principal decision maker is the owner) and does not have any ownership connections to larger and well established firms. Therefore, franchises which are small owner-managed but have the branding of larger and well established firms were excluded from this study.
2. The definitions for failure are outside the scope of this work. However, it is important to note that this is one of the areas of controversy in the literate on corporate failure. Most works define survival/failure using profitability, sales, and liquidation. See Hall (1995).
3. BRICS refer to the fast growing developing economies of Brazil, Russia, India, China, and South Africa.
4. Detailed transcribing: conversation analysis type of transcription was provided. The researcher did not clean the transcript as the intention is to recount the entire narrative for evidential purposes as well. When "..." is seen in the reported narrative in the cases, it represents other things that were said that are not germane to the particular point under consideration.

2
Spur Tree Spices

Abstract: *The case of Spur Tree Spices provides a detailed description of how a small firm that manufactures sauces and spices in Jamaica is able to survive and prosper alongside larger and multinational firms that offer a similar product in the marketplace. The case traces the early beginnings of the firm, its expansion over time and also takes a look at its future. In addition, it focuses on the firm's products, production processes, supply chain issues and quality control processes as well. Similarly, the case also looks at the governance of the business and the business environment within which the firm operates and extrapolates how this environment will impact on its future.*

Williams, Densil A. *Competing against Multinationals in Emerging Markets: Case Studies of SMEs in the Manufacturing Sector.* Basingstoke: Palgrave Macmillan, 2015. DOI: 10.1057/9781137500328.0005.

> Burger King taught me how to run a proper business. You know, how it runs, what makes it work
>
> Mohan Jagnarine, Managing Partner, Spur Tree Spices.

Introduction

Spur Tree Spices is not located in Manchester nor St. Elizabeth, the two parishes in Jamaica which shares the famous Spur Tree Hill. Instead, it is located in Kingston, Jamaica, which does not boast a mountain, but is more known for the Liguanea Plains. Having sat down to speak with the principals of this very impressive small business, which has continued to undertake manufacturing in an economy that is not the most hospitable to manufacturing business, it became clear why the principals have decided to stay the course. Their commitment to Jamaica and their belief in wealth creation were evident from their passion about their business and its future. It is interesting to note that the principals are all non-Jamaicans, who have seen potential in the Jamaican economy and have decided to exploit those opportunities as any good entrepreneur would do. The case reflects a sense of personal determination, serendipity, and family and friendship trumping bureaucratic inefficiencies.

Country information

Jamaica is an English-speaking island in the western section of the Caribbean Sea. The country, with a population of approximately 2.7 million, has more than 52 per cent of its population residing in urban centres today. This rural to urban migration, coupled with other economic woes such as a high debt to gross domestic profit (GDP) ratio, has played a role in the high levels of unemployment. Additionally, more than half of the country's foreign exchange has been gained through areas such as remittances, tourism, and mining. As it relates to imports, there has been a reliance on supplies of oil, food, and consumer goods, which would the level of vulnerability to exogenous shocks.

Another of the issues facing the country has been decades of low to no growth in the productive sector. The manufacturing sector has been at its lowest due to a number of factors which inhibit its growth potential. Chief among them is the country's overall competitiveness ranking when compared to other nations. According to recent versions of the Global Competitiveness Report, Jamaica has ranked poorly in some critical indicators of relating to business competitiveness, productivity, and the environment within which business is conducted. As of 2013, Jamaica was ranked 90th in its ease of doing business, which was three places below its 2012 ranking. While there were some improvements in areas such as ease of starting a business and paying taxes, there was some regression in areas such as getting electricity, getting credit, and trading across borders.

The Jamaican processed foods industry, of which the spices and sauces industry is a sub-set, forms a significant part of the Jamaica's gross domestic product, accounting for 10.8 per cent in 2007.[1] A *Business Observer* report noted that as of 2008, the spices and sauces sub-sector accounted for 2.5 per cent of the total food-processing export, while the entire sector, including value added in Jamaica was close to US$600 million.[2] The share of the exports market in 2008 amounted to about US$10.5 million. This is a part of a larger sector globally which totalled approximately US$8.2 billion in 2009, with the main markets being the United Kingdom (10.3%), the United States (9%), France (6.9%), Germany (5.6%), and Canada (5.1%) being the largest importers.[3]

The case

Origins and growth of the company

Spur Tree Spices officially started in December 2005. It evolved from a partnership between Mr Mohan Jagnarine (A Guyanese-Jamaican who currently lives in Jamaica with his wife and children) and Anand James of similar background to be later joined by Dennis Hawkins (A European-Canadian who worked in Jamaica). The formation of the company was a culmination of years of challenges, trials, experiments, and experiences in the food services industry. Mr Jagnarine has been involved in the food industry in different companies and capacities both locally and internationally. Most notably, he has held the position of Operations Manager with fast food chains such as Chicken Supreme now called Island Grill and Burger King locally as well as with US-based Golden Krust[4] franchise. Mohan and Anand along with Shivnarine Chanderpaul and two other investors were also the owner of De Windies Grill in Mandeville, Jamaica. Mr Jagnarine served as the Managing Director at the restaurant from whence the Spur Tree Spices idea evolved. Mr Hawkins has British and Canadian citizenships and worked for over 14 years in the cash and carry business in the United Kingdom upon completing his studies. He has also worked extensively with T. Geddes Grant in Jamaica as a consultant overseeing their attempts to establish cash and carry type operations in Jamaica since 1974. At the time of writing the case, he had over seven years' experience of owning his own business in Canada after leaving the cash and carry business. The friendship formed at Chicken Supreme, then Island Grill was useful to help the men decide on forming their own operations. What was also helpful is the role that Burger King played in grooming Mohan into becoming a manager. He remarked that:

> Burger King taught me how to run a proper business. You know, how it runs, what makes it work.

With their strong determination and will to succeed, Mohan was determined to not let his lack of academic qualifications stop his progress. Mr Jagnarine revealed that his highest level of education, which he received while in Guyana, was at the GCE O Level. However, his experiences from his very first business, which was a fast food mobile truck called Ranni Fast Food, then to Chicken Supreme, which became Island Grill, Burger King, Golden Krust, De Islands Restaurant in New York, De Windies Restaurant in Mandeville and Sugar Cane Restaurant in Kingston are enough to qualify him as a PhD in the food business. Mr James was the Chairman and Managing Director of Bush Boake Allen and later owner and Managing Director of Caribbean Flavours and Fragrances, a business that manufactured food flavours for some of the largest food and beverage companies in Jamaica. He is also a member of the Institute of Food Technologists. Mr Hawkins on the other hand, completed a degree in French from the University of Southampton, in the United Kingdom. This combination has worked well as a complement in the running of the business. The partners emphasised the business-related experiences and certifications they have been able to garner throughout their working lives. Mr Jagnarine also revealed that throughout the course of his working life, he has also completed management-level and food safety courses, which have added to his own experiences.

Despite the experiences at both Island Grill and Burger King, the foray into business for Mohan really came out of a push factor than merely sitting and planning that this is the way to go. The interesting tale of his journey to New York tells the story.

So I went up to New York for about 2 years ... that is, 2002 I think it was. I went up to New York and worked for Golden Krust at their operations level. But visiting New York and going to live in New York is two totally different pictures. When I went to New York I said no Mohan you remember, you ... by that time my two daughters were attending Immaculate Conception High School. I said there was no way I are going to take out my kids out of Immaculate and bring them in this school environment. This will destroy them. So I told my wife, I say, 'no, stay put. I am coming back home.' I just didn't like it. I just didn't like it, I couldn't adjust. I couldn't adjust. Funny enough, I had been to New York at least ten times before, but as I said, visiting and living there are two different things. I am a diehard Caribbean man. I like to feel the sunshine on my back. I love open air bars and restaurants. For me it was either I go back to Guyana or come back to Jamaica. Because by then I realised that my future lies in the Caribbean.

What I learnt ... because I was in operations, I had to drive all over the place, from Brooklyn to the Bronx and I can appreciate why consistency and convenience in food preparation was so critical. Eventually I made the decision to return to Jamaica and to open my own restaurant. I recall visiting various locations with my friend and business partner Anand James; later we were joined by Shiv and two other friends. We finally settled on Mandeville in Manchester. Eventually myself, Anand and Shiv sold our shares to the other partners.

Interestingly, it was while customers were commenting very positively on our great jerk chicken that we got the idea of creating our own brand. I have a cousin, who has a restaurant that I had helped to set up in Flatlands Avenue, Brooklyn, New York called the Island. Similar Island Grill type of concept. So he wanted to get to get seasonings.

I was making seasonings and shipping it to the States. In the kitchen and the restaurant with two blenders. I was shipping to him through Air Jamaica, in small quantities. By then we had a very good formula. In 2005, the Jamaica Manufacturers Association was invited to participate in the Canadian Manufacturers Association annual fair at a down town hotel in Toronto. Myself and Anand went with some serious samples. We did some jerk chicken and pork at my cousin's restaurant and our booth at the exhibition was a hit! I recall the Canadian Minister of Commerce and Mr and Mrs Bruce Golding commenting very positively on our jerk and pepper sauces. Immediately after that Dennis, who was living in Canada at that time, joined us as our third partner. Since then myself and Dennis have really borne the burden of running the operations while the other partners have lend their support. We have recently been joined by a former colleague at Island Grill, in the person of Mr Albert Bailey.

Products, production, and quality control

Spur Tree Spices specialises in the production of sauces and spices that are used to add flavours to foods. They make sauces and spices for meat and fish products using locally grown Jamaican herbs and spices. The process involved in the production of the sauces and spices is as follows:

1. The farmers bring the peppers, escallion, and thyme to the factory.
2. These are sorted for quality and grading by staff members. This process is done manually.
3. After sorting, the inputs are washed by the workers and then moved to a boiler station.
4. At the boiler station, the inputs are placed in the main boiler to be cooked.
5. After cooking, the herbs and spices are then put into a blender to be crushed and additional material (e.g. vinegar) added to complete the final product.
6. When the final product is done, the output is then moved to a container where bottling takes place.
7. After the product is bottled, labelling is done manually and then they are packaged for distribution.
8. A few of the processes have been automated, including bottling and labelling.

This mix of automation and manual work also helps in the quality control process. While the firm is not ISO certified, the processes are well documented and are verified through constant audits carried out to ensure quality. A robust quality checking system is done by staff to ensure that the highest grade of products enter into the production process. There is also training for staff from a certified food processor in the US, who brings a wealth of experience to the business. Training from the Bureau of Standards also helps to ensure high quality output from staff members as well.

Business expansion and growth

This growth in the business could not have come however without proper network and contacts. Indeed, support from larger businesses is always relevant to smaller business growth and expansion.

One of our first orders came through a friend at Grace Food Services Division. Her name is Tamara Garrell. So Tamara said to me, "Let's put in a bid at some restaurants that use jerk seasonings." Same time the bidding came up and I gave her a price. She put in the bid and the people accepted the bid. We had our equipment stored at Peppers in St. Elizabeth but the landlord could not give us a time frame for electricity in the building. We rented a flatbed truck and brought the big industrial blender to the factory at Caribbean Flavours premises on Spanish Town Road. Anand and myself immediately secured an 80 square feet enclosed area under the carport, got 220 volts electricity in and secured workers to peel garlic and onions etc. That is how we filled our first major order of 100 gallons of jerk seasonings. After this production run we moved operations to Maxfield Park then Woodglen Drive and today a 15,000 square feet plant at Marcus Garvey Drive. (Mohan Jagnarine)

Further market expansion

Grace Kennedy provided some key assistance to Spur Tree Spices, which enabled the company to start the building out of its business. The brand then became recognised and further market expansion was contemplated. This was not an easy road as there were many obstacles along the way. However, with determination and the will to succeed, market entry was assured. The narratives below tell the story of how further market expansion for Spur Tree Spices was secured. Sheer luck and perseverance are key components of this expansion.

> we got onto a distributor just out of sheer luck. A distributor, who was looking for some Caribbean products to sell in a chain called ShopRite. ShopRite is a big supermarket chain on the. East Coast of the USA ... the buyer was there when I met with him, he said to me to go look at some stores. When I went I saw some stores doing rotisserie chicken, so I said to him, "Fred, tell me something we can't do jerk rotisserie or anything?" He said, "Why not?, let's try it, let me ask them." So I said if we can do jerk rotisserie here it would be welcomed and they give us advice to do this jerk rotisserie, and that is how we got into the ShopRite to do jerk rotisserie.
>
> There is a large cash and carry chain, called Jetro, where many restaurants secured their food ingredients and other supplies. Most restaurants buy from big chain stores. My dream was to sell to one of these chain stores. So to get onto the girl is not easy, she is a buyer. You might try everything that you will and get on. So they gave me her number ... normally in the [United] States those buyer don't pick up their phone. They just leave you a voicemail, and if they want they call you back. I tried a whole heap a time!!! One morning ...

> I will never forget it ... one morning about 8 o'clock, I picked up the phone and I called, and she answered and I said, "can I make an appointment?" She must have been in a good mood that day. She said to me, can I come Friday. I said, "if you say come tomorrow, I will come tomorrow." I jumped on the plane, and I went up and saw her. I just wanted to see her, just to make sure to see her.

Having made the necessary contacts and having that important call returned, it was time for Spur Tree Spices to prove that they could deliver on their commitments. Getting the goods to the US was the first hurdle to overcome. Again, friends and networks came to the rescue.

> Luckily I have a friend who was in New York, he was buying some Jamaican products all the time, we are good friends ... we are still very good friends. He is out of Guyana who buys all the Guyana products and export to the US, from Guyana. And he owns a business in Maryland, a supermarket in Maryland ... Jamaican beef. He said he wanted to load a container here ... so I asked him ... man, "can you help me carry up two palettes." So the container for this friend contained his purchases and a few palettes of our seasonings. By the way this friend is now a shareholder in Spur Tree Spices. So he carried two or three containers, and that was how we were able to supply the New York market in the early days. Then from ... there ... we just got into Restaurant Depot.

Getting into the supermarket chain is one step into the wider US market. The next big step is to get persons to buy the product when they see them on the supermarket shelf. Again, the commitment of the principals to see this process through speaks volumes about their business acumen.

> So we made a commitment to Restaurant Depot that they will put our products on the shelves and we would work the market (sampling, demo, promotion etc.) to get the products out of the shelves. In other words, you stock our products and we will help to sell the products. So myself and Dennis went up, I borrowed my cousin's car and I want this to be specific, because I really want people to understand. I borrowed my cousin Ned's car, at that time there was no US credit card, we had no credit card. Stayed at my cousin home, "kotch" one night here, till they get fed up of me and then kotch a next one there ... spend a next night there and start the demo ... myself and Dennis went and demo each store. I went ... he went some of the times and most of the time I engaged Dennis so he had to be here running here. So I am there demoing at these stores, every Restaurant Depot demoing. We borrow a car, you run there, you ketch there.

Self-belief in the product which is sold helps to provide small firms with a stronger competitive advantage as well. This is one of the lessons that small business owners will have to take on board in trying to reach markets for their products.

Business profitability and strategy

Growth in a small business is not instantaneous and can be quite costly for the principals. A lot of sacrifices must be made in order to ensure solid growth for the firm.

> The first three years we made nothing out of it. It was hard work. Whatever we made in profits went back into inventory or new equipment or something to do with the factory ... Growing it becomes harder because each time the orders grow you have to find more raw materials for that order and get more in for the next order, so the accounts will be playing catch up in terms of working capital.

Making the right connection to one's prospective client is critical in the early stages of the business. This will help to get the buy-in into the business as soon as is possible.

> When we do a presentation to a new distributor or someone who is going to buy product from us, we generally marinade the meats with our products and cook these for their staff; feed the office, anyone who is around.

Critically also, is the importance of the brand in helping to grow the business. Small business owners have to believe in their brand and evangelise that belief. In the case of Spur Tree Spices, this was quite evident.

> But you also wear the brand. We learnt this, everywhere you go, brand yourself. Whether it starts conversation on airplanes, in supermarkets, in banks, everywhere you go, it starts the conversation. You don't have big marketing funds when you start off, so you have to believe in the brand, you have to wear the brand, and show how enthusiastic you are.

Similarly, for business growth, diversification of the market is essential. It is critical that small businesses do not rely solely on the markets that they are familiar with when doing international business (e.g. the Diaspora market) but diversify into main stream markets as well. This realisation has helped Spur Tree Spices to grow their markets.

> The other thing we've done over the period is not simply targeting only the Diaspora in the major con-urban areas of the US or Toronto or the UK; because that is a shrinking market to some extent. It's a very price driven market, it's overly competitive. There are brands of everybody in there.

But the vast continent of North America is looking for taste, new flavour, new seasonings. Jerk is becoming mainstream; people want to try things. You have Food Channel pumping out every week for the last ten years programmes on food.

Corporate governance

Complementary skill sets of the principals of the business have definitely helped with the success of Spur Tree Spices. Each partner has strengths in different areas of the operation and these have been put to good use in running a tight operation.

> Why this partnership must work, why it has worked, is because me and Dennis are two totally different people... we work well together, we have major differences but we were able to work out the differences in a way that sometimes we will shout at each other, cry at each other and really get at each other ... but at the end of the day we will then reflect back and then we will say look I was wrong and next morning we move along together.

Taking ownership of the functions in the business is hardly ever going to escape the owners of a small firm. The organisational structure is not always well defined and as such, the owners have to make tremendous sacrifices to ensure the growth and survival of the firm.

> we have disagreements. We end up usually being able to find our way around them and cooling down, and sorting it out. But we have actually carried all the management functions of this company from where it started to where it is now on our two separate shoulders. We don't have the luxury of a department to do this, a person in charge of that, we have no logistics manager, we don't have a production planner per say we don't have ... this is all carried between us. We multi-task. We don't have a creative designer; we don't have a marketing department. These are things we do ourselves and we know our strengths and weaknesses.

The business environment

The dependence on state bureaucracy to support the growth of the small business sector is sometimes not the most hospitable. This was observed very clearly by the principals of Spur Tree Spices.

> One thing we learnt over the period is that you have to do it yourself; don't look to government or government agency to be of any real help to you and

don't look for them to be facilitators, do it yourself, find your own cash do your own thing.

We applied to the government for factory space. The previous factory space was inadequate and the landlord was only interested in maximising returns. I call the Factories Corporation and tried to negotiate a rental agreement for our business. The Board flatly rejected us and seemed to be more interested in how much US dollars they can collect in rent up front. We asked for a reasonable rent coming in to use our capital to create a modern factory then we can pay an increased rental. They write back to say the Board can't accept that. I wake up one morning, call the Factories Corporation again. One thing I always do when I call, I ask the people I want you to come and see physically where I am. We got some people from the Factories Corporation to come to Woodglen Drive see what we were doing there. They visited and we started to get somewhere with them.

It was really the most frustrating period in my business career. Let me tell you what happen one day. I picked up the phone and called the JMA[5] and I was really emotional I spoke to Metry Seaga. I said Metry, you know what I am going to do, me as a Guyanese a go block the road, me a block de road. We want justice too. And Metry said in a board meeting he mentioned, that when he heard my voice; he heard the cry in my voice he sense that my business was in trouble. He said, he heard, "you are getting nowhere, you are getting nowhere". He took up the issue and went up to the Prime Minister but we even wrote [The Honourable Anthony] Hylton[6] a letter. Met Hylton and ask him to dispatch a letter tomorrow morning up to now we don't get a reply. We write Minister Hylton saying this is what we are doing, this is what we are doing, this is legal, intervene and what have you and so on.

This episode, which reflects the difficulty in acquiring a location, shows how challenging doing business in Jamaica can be for small enterprises.

Besides the difficulty of doing business in the domestic environment, Spur Tree Spices also faced serious challenges in its export business.

> The ease of doing business as an exporter is crazy. Do you realise that if I wanted to send this jar of jerk seasoning to a customer through UPS or one of the couriers, FEDEX, I have to go and get a letter from the Bureau of Standards, saying I don't need a certificate of export as an exempt item. Now that will go and cost me a $1000 and it will take me at least two days to get... Unbelievable!!!

The future of the business

Having moved from a blender in the back of a kitchen in Mandeville to becoming a US$1 million export business and accessing markets in the

US, there is no doubt that Mohan and Dennis will not rest until they see the full potential of their business.

> put it this way. Where we are now, we can't continue along the lines we are going. That is why we are trying to bring in an equity partner. We have to find the management skills, we have to get a right production manager.
>
> we want to eventually have somebody who can grow into a general manager. You will never able to ... do that without putting a plan together. Because ... to do that, you have to increase the business to find more business to pay the person etc.
>
> But we've reached the stage where we realise that and Albert Bailey has taken the lead as our CEO to get the systems in place to facilitate expansion and growth; including financial management, accounting management and staff management, production management, and planning in terms of production and planning in terms of export marketing.

The thinking is clear as it relates to the level of professionalism which these partners want to instil in their small business. Not to be left out is the quality of staff they hire and the continuous upgrading of staff that they believe will be able to assist the firm to deliver a strong performance. Unlike some small businesses, Spur Tree Spices is keen on investing in training and development of its staff.

> Some are being sent on courses to the Bureau of Standards. Just recently we had a PhD in Food Science from the University of Alaska training our workers through the friendship I have with his father who is a retired professor in food science. He did not charge us. He is also the head of the certification body in Alaska. He came down last week, he took all our people through the extra basic training, looking at what we are doing and making recommendation and training the staff hands on. I have a young lady here who is doing the training as well herself.

Business and strategic planning

In addition to staff training, the management team is also seriously focusing on designing a long-term strategic plan to move the business to a higher level. They recognised that merely focusing on operational issues will not be sufficient to move the business forward. Direct efforts at strategic planning will be needed. The management is aware of this, and as such, is making the necessary preparations to have a good plan in place.

we have put together the plan. We are finalising it now, it is partly to use in terms of soliciting additional investors and put in investment but also from our own point of view, to formalise where we are going and to put us on a different structure for the next five years. We expect to more than double in the next five years, and that's just in our existing markets. We are following up a large number of "sighting" opportunities in the flavouring markets in the seasoning markets for processed foods. We have talks with a chicken plant in Canada about supplying rotisserie chicken to the largest supermarket chain in Canada. We already have a programme on the way with ShopRite stores, which can be expanded and if we can get a chicken supplier there, we hope to send in our ready marinated products, injecting the marinade again that will open up that market.

One of the reassuring signs that the business intends to continue to survive is this quote from one of the partners:

And we are looking all the time, new markets, new products, new ways of using the skills we have. You can't stand still, you can't sit still either. You have to innovate; you have to think outside the box in terms of everything. If you don't do that you are going to be dead in the water.

This is indeed a strong signal that the partners understand the highly competitive space in which they operate and the need for continuous improvements in order to ensure their survival in the market place.

Concluding thoughts

Spur Tree Spices moving from a mere concept at the back of a kitchen, where one of the partners was making seasoning and selling it informally, to becoming a US$1 million export business is not an effortless task. The manufacturing sector in Jamaica is tough to operate in as seen from the many obstacles faced by the company. However, despite this less than hospitable environment in which the firm operates, the principals along with a dedicated staff have found strategies and tactics to overcome the difficulties to ensure the firm's survival.

Critical to the success of the firm so far is the adroit leadership displayed by the main principals Mohan Jagnarine and Dennis Hawkins. Now joined by Albert Bailey, their perseverance and their belief in their brand; the role of networks and friendship in helping the partners to achieve their dream, and the dedication of their staff will further enhance the development of the business.

the one thing I have learned and I have been involved with a number of companies, and a number of my own businesses and they have been in this group/field... but the one common trend is you have to have common sense in business. And you have to get a good team around you. (Mohan Jagnarine)

These factors have all helped to ensure that Spur Tree Spices will continue to survive and prosper in a tough manufacturing sector that is characterised by heavy bureaucracy and an inhospitable domestic and international business environment.

Notes

1. Market Brief Spices and Sauces: Jamaica, 2008. Jamaica Exporters Association.
2. *Business Observer* article entitled, "Local Spices, Sauces Present Strong Growth Possibilities" – 6 July 2011.
3. Ibid.
4. Island Grill, formerly Chicken Supreme, is a wholly owned local restaurant chain owned by Mrs Thalia Lyn. The company was started in 1991 and currently has 18 outlets. Burger King is a part of an international franchise with 28 locations across the island. The chain has been locally operated by Restaurant Associates Limited since 1985. Golden Krust Caribbean Bakery and Grill is an American-based manufacturer and franchisor whose owner has Jamaican roots. The company has over 100 restaurants in nine states across the United States.
5. Jamaica Manufacturers Association
6. Hon. G. Anthony Hylton is the Minister of Industry, Investment and Commerce in Jamaica.

3
Yono Industries

Abstract: *The case of Yono Industries traces the development of a small detergent manufacturing plant in the manufacturing sector in Jamaica over the years to better understand the factors that contribute to its survival and prosperity over time. The case traces the early beginnings of the firm, its expansion over time and provides insights into the future of the enterprise. In addition, it focuses on the firm's products, production processes, supply chain issues, and quality control processes as well. Similarly, the case also looks at the governance of the business and the business environment within which the firm operates and extrapolates how this environment will impact on its future.*

Williams, Densil A. *Competing against Multinationals in Emerging Markets: Case Studies of SMEs in the Manufacturing Sector.* Basingstoke: Palgrave Macmillan, 2015. DOI: 10.1057/9781137500328.0006.

> We have a fully vertically integrated business... from ideas all the way through complete finished product from designing, packaging; for instance we can make our own bottles, we can make our own caps, we work with any kind of plastic. So we do blow molding, injection molding, stretch blow molding; we can work with any kind of plastic resins.
>
> Dr Andre Jones, Managing Director.

Introduction

The creativity and entrepreneurial drive of a Jamaican-American, led to the establishment of Yono Industries Limited situated in rural St. Andrew. Yono is indeed a very interesting manufacturing plant with its owner and principal playing a significant role – from conceptualiser to executioner of almost all business processes. The company is fully vertically integrated from ideas to the completed product that is sold to the final consumer. Being one of a few players in the fragrance section of the manufacturing sector, Yono has done an incredible job to remain open in an environment that is less than benign towards manufacturing. The high cost of electricity, the increased competition from Chinese imports, the unstable currency, the high cost of security among other impediments are all factors that could have turned Yono away from doing business in Jamaica. However, the entrepreneur had decided to commit to his vision of building a world-class manufacturing business domiciled in Jamaica.

Sub-sector information

Information on the global fragrance industry indicates that this segment is worth between US$8–10 billion according to the International Fragrance Association estimates. The division of the market sees the household section of the industry accounting for 49 per cent, personal care at 25 per cent, fine fragrances at 21 per cent and others accounting for the remaining 5 per cent of the market. As it relates to sales, North America (34%), Western Europe (28%), Asia Pacific (24%), with Eastern Europe (2%), South America (6%), and the Middle East (6%) accounting for the remaining percentage. Conservative estimates reveal that growth rates have averaged between 2–3 per cent over the past decade.

The case

Business origins and expansion

Yono Industries is a research, manufacturing and development company that was founded by Dr Andre Jones in the United States in 1999. The creativity of the man is manifested even in the name of the company. Yono got its name from the owner's unique style of thinking and execution, and his long-term vision for his business.

> Well, Yono came about because my sons are named after a tennis player Yonnick Noah. So my first son is Yannick and the second is Noah, and so I took the first two letters of their names and called it Yono...I wanted to create something for my children, a legacy, and I wanted to hand this company to them, and so that's how I came up with Yono.

The company was originally slated to be set up in Jamaica; however, there was an absence of follow-through on the part of the individuals who he had consulted, when he had first decided to start the company. However, the bureaucracy and difficulty of getting things done in an efficient manner in Jamaica forced the owner to resort to setting up the operation in the US. The entrepreneur noted that it was always easier to set up a business in America but he always had in the back of his mind, the inclination to establish some form of operation in Jamaica.

> So in fact the company that I am currently operating under, which is Yono Industries, is actually registered before my American company, which is Yono Corporation. And... but I still was intending to come back home; really wanted to come back home to Jamaica.

The Jamaican operation was to be opened much later in the life of the corporation. The owner of Yono Industries is a trained chemist by profession. It is not coincidental therefore that this line of business finds favour with him. He has experience in various aspects of chemistry; including but not limited to synthesis and pharmaceuticals. While familial influence was important to push him into business, although he is a trained chemist; the internal entrepreneurial drive was always there from a tender age.

> I realised from I was very young I love to build things.

This realisation led him to think about entrepreneurship as a way of breaking the dependence of working for others and also to help to build his native Jamaica.

> Currently, I still live in the US for about 26 years now, so I have been living there from 1986. I left [after] high school [in] Jamaica [and] migrated to California to complete my tertiary education. I started there, built the company and still was looking to come back home. Actually I had several different conversations with different people in the US about young Jamaicans coming back to Jamaica, to help build the country, and I decided that I would pretty much take them up on the offer, and move my entire manufacturing plant back to Jamaica.

Interestingly, although the desire was always there for business, it was not until his doctoral work in the field of chemistry that he really cemented the idea of starting his own company. The independence he gained from his PhD studies showed him that he could work on his own initiative, and therefore, it motivated him to start his own business instead of working for others. He clearly did not have the desire to work for one of the larger research and development (R&D) companies, which was fashionable for most of the colleagues he encountered at university in his PhD programme. So intense was his desire to have a business, he began conversations with his professors and fellow students about owning a business. Being intrigued by this desire, they questioned his interest in completing a degree in the sciences as opposed to a MBA. He however was not keen on further studies in business as he did not believe he needed to spend any more time in the formal education system. His sole focus was on completing his doctoral work and to start using the knowledge he gained to make money. He revealed that it was then that he started working on establishing his company.

> I want to be free where I can travel, go anywhere in the world, set up my company and do what I want to do without anybody controlling me.
>
> So before I even finished school, I started putting together the business plan understanding more about the business, so I didn't really go to business school, I didn't have time for it. I remember one of my professors saying, you talk so much about business Andre, why you don't go get your MBA, and I say I don't have time; I just wanted to get my PhD and get out of here. But I was literally designing and planning the business in graduate school and one of the big questions was, how am I going to raise sufficient capital to start the business. And so...you know, I could build a spreadsheet look at all your costing and so on, to see where I need to come up with the money.

So I started talking with different individuals. Before I even finished graduate school I was able to sell US $50,000 worth of shares in a company that I literally had on paper and whatever other savings, and stock holdings I had, I sold the rest of that and started...before I even went and work for somebody. And so I started a company in Florida, which was Yono Corporation and when I ran out of cash, cause I had to buy equipment and all that other stuff; and started run out of cash for a little bit, and I went and work for Watson's Pharmaceuticals but only to earn enough money and also to close a contract.

So while I was working for them doing...as a QC Chemist doing--, analysing OxyContin, Traumadol, all those narcotic analgesics {up there}. I was negotiating on the side trying to make hotel amenities for soaps, for shampoos, for different stuff like that and when the contract came through it was a couple of hundred thousand US. I resigned and went back fulltime into running my own company.

One of the true tests of the entrepreneurial journey is to be a maverick and go against family tradition. This is not uncommon in an entrepreneur's life.

It was hard to convince my own family that I wasn't wasting time, cause my mother think that I should just go get a job and I was just being lazy; (laughs) but it's not being lazy I said, listen you can't make money ... what I sell. I make more money in a month than what that company probably pay me for the entire year if I close this right deal or that deal. They don't understand because she might not necessarily have the track record of business and how money, real wealth {makes it}. I was interested in real wealth and how it works. I said money in itself is not wealth. Because I have seen where my mother, when she left this country left hundreds of thousands of dollars in a bank down here and it became worthless; inflation eat it off. Bank don't pay interest, this, that and all kind of stuff. I am explaining all those things to her but I didn't have a track record of being a financial mogul or even a finance degree but I could tell her it's the truth.

Products, production, and quality control mechanisms

Yono Industries is a fully vertically integrated company, which manufactures and sells fragrance and other personal care items to retail and wholesale customers in the Jamaican market.

we are fully vertically integrated from ideas all the way through complete finished product from designing, packaging; for instance we can make our own bottles, we can make own caps, we work with any kind of plastic. So we do blow molding, injection molding, stretch blow molding; can work with any kind of plastic resins.

The company produces lotions, perfumes, shampoos, body wash, soaps, and a host of other body care products used for daily hygiene. Overall, the company produces a range of over two dozen hygiene products.

> We make products for other company. I have made hotel amenities for Super Clubs[1] while I was in United States. I used to make pure hotel amenities, such as shampoos, hair conditioners, mouthwashes, shower gels, soaps.

The company does not only manufacture these items, it also manufactures the package and produces the labels, which are placed on the products for distribution. This is part of the vertical integration that the owner refers to.

> we can make our own bottles, we can make own caps, we work with any kind of plastic. So we do blow molding, injection molding, stretch blow molding; can work with any kind of plastic resins.
> We make bottles, we make caps, we can do printing, we have an entire printing division.
> Basically, when you see bottles, they are made from a two-step process. So, from the resin to the pre-form, from the pre-form to this [the bottle]. We are one of two companies in the entire Jamaica that make pre-form. ... we make pre-form, most people import those, we make them.

Besides using its own containers for packaging, the company also makes its own labels and does its own printing as well.

> We buy labels and we also can print our own. So we are actually getting ready right now to print appliance products. So we are actually trying to do some training to young persons who have got graphic designing experience but not really much in terms of printing. I am teaching them from going from designing on a computer to how to output negative or positive and making printing screens or edge printing plates and so on.
> And for screen printing, this is the machine that we are setting up right now to do all the printing on the back of the packages. And these are just more screen printing machines.

The production process

Yono gets a competitive advantage in the production of its final products, mainly because it sets up a production system that is completely dependent on alternative energy and does not depend on the expensive electricity from the JPS grid. The cost of electricity in Jamaica has been a prohibitive factor for small manufacturers. There is a consistent call from manufacturers for the state to do something about the cost of energy in order to help to improve their (manufacturers') competitiveness. Yono, being the proactive company that it is, did not wait on the state to reduce the price of energy but instead went about designing an alternative system that could deliver lower cost energy to them.

> I generate my own power... This is our hybrid system, these are 500amps B {cycle} batteries so we have about 34 of these hooked into an inverter system where we can generate up to a 6kw.
>
> Basically, I use a big generator when I am using a lot of large power to run like the big machines, and then any excess power I am not using at that time get stored into a hybrid inverter system, a battery system that we will pull back power from. We run all our lights, run smaller machines like we need. Fully off the grid but we are able to manage the cost.

Similar to its electricity generation, Yono also purifies its own water. The company has to store water for its production due to the unpredictability in supplies. Additionally, due to the company's strict adherence to providing quality products, the company has its own water purification processes, which ensure that the water that is used in the manufacturing process is of the highest quality.

> so we also purify our water both for profits control, to control temperatures of the machine, the processing temperature as well as we use it inside the product to make the different products.

Yono does not have full control over the raw materials it uses for some of its products. It generally buys material from China or from local producers. However, the owner laments the haphazard delivery of raw materials from local suppliers; and as such, he depends heavily on imported material from China or the US.

> We import most of our raw materials into Jamaica for reliability of supplies. There are companies that supply chemicals here but they are importing it from the same people that I would import from out of the US or out of China. For the resins, plastic resins, I import out of China because it's more reliable... Raw materials that we use in making shampoos, we are talking about different types of surfactants and emollients and so on. We import those out of the US.

Product development takes place in the same outlet where production is done. The product development is mainly done by the principal owner, Dr Jones. Given his background in chemistry, he spends a lot of time in the laboratory, which is located on site; thinking about and designing new products for his operation.

> This is my research lab, where pretty much I developed the products that I work on... from ideas, you run small scale batches and you go down, you go down ... I am the only R&D Chemist.

Indeed, the R&D at Yono is motivated by the owner's own thinking and experimentation or the solicitation from clients.

> a client might tell me they want this particular type of product to get into the market. They can identify what category of the market they want, if they want a medium price or a low cost product. I could come in here, develop the product, do some cost analysis to see where/what the cost structure of the product would look like; the product's table, we say, yes we can take it the market, what type of packaging they want to utilise. So we can do all of that product line or for myself.

Yono has an elaborate production plant that is befitting of any first-world operation. All the machines have an impeccable look of newness and are well laid out in preparation for production. The production flow is logically laid out and easy to follow, from raw material acquisition to packaging and delivery. The plant, although located in a small factory space, in terms of its equipment, can be described as world-class. This seems to have emanated from the owner importing all the equipment from his base in the US, to set up the Jamaican operation.

Corporate governance

Similar to most small businesses where the owner is the main point of contact, Yono Industries is no different. The owner, who classifies himself as the managing director, carries a heavy workload. He does most of the business planning, strategic thinking, financial management, marketing, and customer attraction. Currently, the company does not have a Board of Directors as the owner does not think it has reached that stage yet.

> The company is set up as a corporation, and there is a reason why you want to have a corporation – you just learn from big companies, you limit the liability to your person.

We don't have a general Board; our Board is at best ... currently, right now, myself and my business partner as the directors. Yes we would like to assemble a Board but I don't believe in putting people on a Board if they not going to bring something to the table and be able to deliver. So I am not going to load up my Board with people just because they have titles, and they have names and they are not bringing anything to the table. They are neither going to be strategic investors or whatever, so I'd rather leave those positions for ... if and when I really need to go public, either listing on the Stock Market, Junior Market or raise massive amount of monies to say supply a potential big client that they have a say in coming on board, and putting their directors, so I leave those positions open.

The Yono Corporation is a partnership between Dr Jones and his business partner who is based in Florida.[2] His business partner (who has a MBA in Finance and is also a Certified Public Accountant) and himself constitute the "Board of Directors". This partner acts more as a silent partner in the business, while Dr Jones takes on the day to day operation. He describes the partner as the President and Chief Financial Officer of the Corporation.

Further, Dr Jones's impatience with rhetoric without action, has delayed his desire to set up an established Board of Directors for the Corporation. The "Board" is limited to him and his business partner.

In our business or in our world amongst black folks, we spend too much time talking, we talk and we don't do nothing. I am usually...when I was in California I was one or two black kids in a PhD programme of 200–300 students, and it wasn't about talk, it was all hands on doing stuff. We talk about business rather than doing business and I am really against that, I think we need to just do stuff.

Customer and marketing

Yono Industries like other small businesses does not employ a large marketing team nor does it have mega marketing campaigns. It sells mainly to wholesale customers who then retail the items. The company also does a bit of retailing on its own. The owner, while employing persons to carry out the marketing and sales function, is quite heavily involved as well.

Usually most of my marketing is usually through personal direct, we call direct marketing. Go and talking to people, not from advertisement or whatever it is. Advertisement work when you are already in the market all over the

DOI: 10.1057/9781137500328.0006

place and just basically want people to recognise your brand. People have to understand the difference between advertising and promotion versus really getting your product into market where people can buy.

We were in maybe about 15–20 stores in Jamaica including the airport, so we were in like the four Things Jamaica stores in the airports that would sell our products to the tourists and so on. And we were like in other pharmacies, some of the major pharmacies in Jamaica. When you are operating in Jamaica you are open to a lot of challenges; you hire either a sales rep and you want to manage it yourself and they go out and sell. We have had that unfortunate incident where people run off with money, run off with products, challenges with collection. So I put in place a marketing manager, who is now going to manage that programme but basically what I do is to sub it out – meaning that you are an independent marketing manager, you are only going to get paid if you collect our money, and turn over the money to us and you see the gains keep on going on.

Yono also sells its products in the export markets using mainly its Internet store. However, the export side of the business has slowed down due mainly to decreased demand.

We do exports but that has slowed down a little bit. I also have an Internet store, when we get orders through the web, we usually ship to Miami, and we sell off the products when they order them through the stores. So we also have built up some web base capability. We have small individuals right now that order our products they resell it in the US. But one of the big potential clients that I was really targeting was Target stores, unfortunately the wreck that was happening in our accounts got {kicked} and it really set us back big time. So now we have to look at {near shore} our own market to get things moving in the way we want to.

However, given the challenges in the local business environment, expansion in the export side of the business is also affected.

It takes money to market, if you understand; there is many ways you can market products. I don't know so much about you but I know about America since I have lived there a very long time. It takes money to market, I can show you a signed Walmart agreement that I got back in 2003. When I got that agreement they gave me six Walmarts to supply. And I needed to expand my facilities and bring on people and I couldn't raise the two or three million dollars I needed at the time.

Unquestionably, Yono prides itself on selling at a competitive price in the market place given the efficiencies it gains from the operation of its factory.

I don't know about others in the business but in my business, I can match prices out of Trinidad or beat them out of China. I can be very competitive. My plant is very automated. I can source raw material out of China if I need to, not always have the quality in some things but you can make those evaluations or sacraments as you go along.

Employee engagement and training

Although the company employs a number of persons mainly on contract, the Managing Director, Dr Jones plays an active role in the supervision and management of the employees. He does not have a separate human resources manager to look after the affairs of the workers. In addition to his many other duties (financial controller, chief scientist, research and development expert, and so on), he also finds the time to supervise employees in the factory.

> Right now I really keep one permanent staff member; everybody else is just temporary; just to manage my cost just to survive until something comes through.
>
> I supervise the entire factory. It's not the best way but what you find is you have to train your supervisor, to oversee your supervisor. People are not well trained here. It's really a serious challenge, we make light of it but it's a serious challenge, it will break you. And luckily that's what saves me because I am so hands on. I am able to do a lot of the troubleshooting, maintenance work on my own equipment, even though I am not a trained engineer but the point is, I have just learnt over the years how to do a lot of stuff. I can work with electrical stuff to a certain extent, mechanical stuff; so I will turn on a handle grinder start fabricating and cutting, and welding and whatever it is, and they are like wow! You are a Chemist, how do you do that? I have hired several of our engineers here out of the University, and I have to fire every single one of them.
>
> I fired every single one of them. Why? Because they make mistakes: for them it's just a mistake while it costs me millions. And until they understand that ... sometimes they don't take you serious, so the only time they take you serious is when you fire them. That's a big challenge for business in this country. If you going to bring a high-tech business, you have to know how you are going to maintain that business or keep it running. It is serious, serious problem.

In addition to supervising the technical works of the factory, the Dr Jones also does all the logistics work for the company. Again, his training during his PhD studies has helped to effectively carry out this task.

> I deal with all the logistics in terms of sourcing of raw materials, those things are not difficult because remember I have been doing it for so long. When we were in graduate school being trained as chemists we ordered our own chemicals. We just tell the professor or whatever it is, put it on this P.O. or they give us a P.O. number, and we just order everything, so we dealing with logistics from ever since.

Staff training and development is very important to the company; so as to ensure that the performance of the workforce is at its optimal level. This is especially so, given the very technical nature of the operations that it runs.

> Training is a critical factor in any business, you have to train people in process, development, how to operate machines...like safely, there is a cost to those things. If we train our young people at HEART say in certain things and it's not related to my field specifically, they can't help me. I gotta do all of that training; that's where the support should come in. I am willing to partner with different institutions to train people but it's gotta be a quid pro quo, simple. I have a teaching background, I thought chemistry both at the University of California Sciences and Harvard University, so I know about teaching and for years, right. But I understand business, so I go out there and do it, not talk it. I am in business almost 15 years.

The challenges of doing business

Dr Jones spared no punches in outlining the various challenges of operating his business in the Jamaican environment. He does not see the policymakers as savvy enough to understand the rudiments of doing business, and as such, they design and implement policies that are inimical to the effective operations of businesses especially small firms.

> And how are we going to move forward as a country if you are not going to have policies that foster growth and development? Seriously, when you have a manufacturing plant you have to hire young engineers, young scientist, business students, marketing students, sales people you are going to train those people. Who is supposed to do that? Are we all going to get a visa and migrate? You have to support the productive sector and you have to also support small and medium business not just the big guys alone getting all the supplies. It can't be just about talk, and if you are not serious as a politician or a policymaker, don't come to people like me in America and tell us come home. Cause my business is not about talk, I don't get paid to talk I get pay to produce.

You have to have targeted investments. You know I have heard the discussion where people are talking about picking winners but guess what you better start picking some winners because you are doing the same thing over and over and coming up zero.

But we spend what...US $80 million goes to one sub-contractor. So you widening now, you've raised it, how many jobs is it creating, is it sustainable? We have to look at sustainable development. Correct? So US $80 million a few guys benefit, ok. But is it sustainable? Is it continuously generating jobs? Can you feed families day in day out of that stuff? The answer is no, let's say you had taken off US $5 million or US $10 million and say, I am going to have some targeted investments. Look at my investment, what it is worth inside this country right now as a young black Jamaican, and you target that put that into a company. What could the turnaround be? What could be the return on investment in terms of human personnel training, in terms of being able to develop and market the product to overseas clients or even supply our local hotels? We have 30,000 hotels rooms in this country with all those have to have hotel amenities. Tourists have to bathe but why do they need a soap imported from America, do they need a toothpaste imported from America? I can make the same toothpaste, the same shampoos, so where is the support? Come on, where is the support, seriously? These are basics in terms of business development. Obama does it for his own company, he does it for GE (General Electric), he does for a lot of other companies, why are we not doing it for our own people, our own country? This is disingenuous, so that's the reality and I think that's how we tell our young business people.

I want to say something about the ease of business in Jamaica, there is no ease of business in Jamaica... There is absolutely no ease of doing business in Jamaica, it is all talk. You're going to lose all you money before you even ... I will tell you the truth, I am not afraid of saying it. Take for instance your EXIM bank, I am a client but right now I have the bank taking between 70 to 82 per cent of my receivables... How you do expect to build a business and grow your business if the bank is undermining you that way?

I have a loan with them; they say it's for $45 million ok. I put up $114 million as collateral. Instead of structuring a loan, where they could give you a longer term so it can work, they say they want to get ex pay, this, that so they sign a tripartite and they taking all the money. That's not right. Don't tell people, come on I am not that old! Don't tell people to go into business and this is what you face. I am saying it, it's nothing hiding, it's the reality. Where is your working capital going to come from? You are gonna go out of business if you can't manoeuvre fast enough.

From the tone of the narrative, it is clear that Dr Jones does not feel inspired by the policies developed to support businesses, especially small ones.

Future of the business

Despite the difficult business environment which Yono finds itself in, the company is still keen on surviving and prospering in the market place.

> I am looking at a couple different options and one of them really has to do with succession planning. I'm getting older and I want to leave a legacy for my children and so one of the things maybe is looking at either going public on the Junior Market or Stock Market here. If that's not viable, then I am also looking at pulling back out of Jamaica because I don't believe we have created the environment for people to be successful in this country. And it is very, very sad but somebody needs to tell the truth and I don't mind telling the truth. We are not creating the environment for people to be successful in this country. I was successful before I came back home to Jamaica. Ok, remember, I left here from high school. I have an advanced degree in chemistry and I was successful, and I made it in America. I nearly became a millionaire in America I can tell you that. US $80 million and come back here and you have to be fighting every single day, that is not right because we are losing our best and brightest but nobody cares because I can tell you, I have to beg access to our policymakers, and I am telling them the truth and if they not doing anything, then I know it's disingenuous.
>
> Not a lot of people have access to our politicians here. I talk to at least a few I'm telling them the truth I even ask them to help facilitate this, facilitate that. I don't know why they don't do anything, that's the reality. If you lose people like me who are so capable and have so much knowledge and know how, so what about the people who have so much less, what are they going to do? What is going to happen to the country? We really need people to be honest and step up to plate. When people tell me don't come to Jamaica, I say, I love Jamaica this is where I born, this is where I grew up, I want to come back home, I can do a lot, I see all the potential – it's all there but is it benefiting us as people – black people, is it benefiting us as a people?. Other people are able to come here and be successful, you know the reason why... they work together as a team, as a group, we are not doing it.

Business planning is also critical to Yono's future success but most importantly is the need to have a practical plan that takes into consideration the realities of daily operations.

> I mean, I have a formal business plan but as I have told young business students even our interns; I have had the opportunity to work with some of UWI MBA students and I have actually showed them how to write a business plan as a scientist but I said that's all good and well, it's on paper. But I show them the mechanics of writing a business plan, how do you {tie} your sales

forecast to your cash flow, your balance sheet, your start-up capital and all that stuff, but that's theory. There are critical assumptions you have to make all day long, and if you miss those little assumptions it's going to make or break it. Look on the business plan it works, look at your financial it works. You can make profits on one paper but if your collection is off or your time for collection, you are broke, you fail; you run out of cash, your cash flow is critical. They have to understand that sometimes you have to feel it, you have to know it.

Concluding thoughts

Yono Industries, despite the challenges of financing, access to high quality workers, overcoming government bureaucracy and red tape, gaining consistent supplies from local market among other crucial issues, is still prospering in the manufacturing sector in Jamaica. There is no doubt that the dedication of the owner to making a difference in his native Jamaica is a strong driving force in the continuity of Yono Industries. The owner has the luxury of moving back to his country of citizenship, the US, where a lot of the challenges he has encountered in the local market can easily be dealt with. However, he has decided to continue to give Jamaica another chance to create employment for the citizens of his native land.

> the business could easily ... if we were getting the support ... hire 25 people with ease because I have enough production line to run constantly. We can't do it unless we get some support. We can't have the bank taking all the money and not giving us proper terms to work with. We can't have the EXIM bank doing that, that's the truth. I don't care if they like me don't like me. Maybe some people think that your success is going to come if somebody likes you – rubbish! We have to compete with the rest of the world and any of my products, I have 24 of my products registered with the FDA; can go through Customs – that's what I was shipping out of here in Jamaica, 24 different products registered. Where is the support?
>
> You cannot say your people are free unless they are economically free. Everything else is talk and you have to have people, and you have to have policymakers who understand production, who understand development, and who believe in it. If they don't believe in it, it will never happen, ok. I am still a resident of the US I can always go back. My knowledge will allow me to move to another country but what about the people in Jamaica? What about the rest of them, that's what we want?

These are indeed very strong and passionate words that both policy-makers at the micro and macro levels will have to heed if Jamaica is to continue to see more of its small manufacturing firms survive and prosper.

Notes

1. Super Clubs was a large hotel chain in Jamaica owned and operated by the Issa's a Lebanese/Jamaican family. At the time of writing this case, the hotel chain was no longer operating in Jamaica.
2. Yono Corporation is the holding company for Yono Industries and other business interest of Dr Jones.

4
Island Moldings

Abstract: *Island Moldings, a small furniture manufacturing company which produces moldings, windows, and doors in St Kitts and Nevis has been surviving and prospering in the tough manufacturing sector in the small country of just over 60,000 people. The case outlines the early beginnings of the firm, its expansion over time and provides insights into the future of the enterprise. In addition, it focuses on the firm's products, production processes, supply chain issues, and quality control processes as well. Similarly, the case also looks at the governance of the business and the business environment within which the firm operates and extrapolates how this environment will impact on its future.*

Williams, Densil A. *Competing against Multinationals in Emerging Markets: Case Studies of SMEs in the Manufacturing Sector.* Basingstoke: Palgrave Macmillan, 2015. DOI: 10.1057/9781137500328.0007.

> One of the things I like to say is that other people may know and you don't necessarily have to have money to start a business because a lot of people may think that you have to have a million dollars to start a business. Island Moldings started with $15,000 and I love to say it because it's something that I think is good to motivate people.
>
> Donald Hendrickson, Owner and Managing Director.

Introduction

Faith, courage, hard-work, and determination are the factors that one could say has brought Donald Hendrickson and Island Moldings, a company that specialises in wood craft (windows, doors, and moldings) for residential homes and businesses to prominence in the small Island of Nevis. The story of Donald Hendrickson, who was a very quiet and reserved child in school, and who later became a major employer of persons in his native Nevis, is not only inspirational but has many life lessons that other entrepreneurs can learn from regarding how to build a successful business in a hard economic climate.

Country profile

The twin-island federation of St Kitts and Nevis has a total area of 101 square miles and a population of 53,584 according to 2013 estimates.[1] The main economic activities in the country have revolved around sugar-processing, tourism, and the financial services. Since the mid-1980s there has been a concerted effort to diversify the local economy away from sugar. Along with tourism and financial services, export-oriented manufacturing have begun to play a larger role in the economy. It is reported that the United States is the country's most dominant trading partner, accounting for 61 per cent of imports and 84 per cent of exports.[2]

As it relates to doing business, the country has dropped four places to 101st in 2014 on the Global Competitiveness Index for ease of doing business. There has also been a decline in the country's ranking in the ease of starting a business (from 67th in 2013 to 73rd in 2014); getting credit (from 126th in 2013 to 130th in 2014); paying taxes (from 142nd in 2013 to 145th in 2014) and protecting investors (from 32nd in 2013 to

34th in 2014). There were some positive indicators however, with the country having a comparatively high ranking in dealing with construction permits (13th in 2014) and getting electricity (19th in 2014), both of which can directly impact on attracting new investments as well as the manufacturing sector.[3]

The construction sector is said to contribute approximately 12.94 per cent of the gross domestic product (GDP) of St Kitts and Nevis in 2012.[4] This figure however, represents a marked decrease in the domestic economy from the height of the decade between 1990 and 2000 when the sector accounted for 19 per cent of the country's GDP.[5] It is driven in large part by the infrastructural investments in the local hotel industry followed by domestic housing programmes and infrastructure. Currently, there are two major companies involved in the trade of furniture in the country; they are the St Kitts and Nevis and Anguilla Trading and Development Company and SL. Horsford and Company Limited. These companies do not only deal in furniture but are also diversified trading companies, which are generally involved in multiple trading, services, and manufacturing activities.

The case

Business origin and expansion

Fifteen years ago (1999) in his home in Nevis with only EC$15,000, Donald Hendrickson had the bright idea of starting a small business to manufacture doors and windows (and later moldings), for household and business places. The seed for this idea was planted during his days in school and subsequent to his graduation with six subjects at the secondary level. However, given his lack of interest in academic work due mainly to his teachers' misinterpreting his quiet demeanour as dumbness, he decided that he would try to do something more practical after leaving secondary school. He went on to work as a gardener at the Montpelier Hotel. It was his first work experience but he recalled the intense heat he felt from the sun and vowed from there that this would not be his final job. He eventually began to like gardening and stayed on at the hotel for two years. He was later fired from this job. This could have been a blessing in disguise.

> I got into gardening – I like it and I was enjoying it at Montpelier Hotel. I worked there for two years, and one day I got fired, you know and I mean,

> according to my Christian faith sometimes God allow things that happen to elevate you. He is not going to give anything to suffer you. And I really felt bad but the amazing thing is that there is a guy who still works at Montpelier doing gardening, who was employed at the same time with me, so if I didn't get fired I might have still been at Montpelier doing gardening because I did enjoy it.

After being forced to leave his gardening job at the hotel, Mr Hendrickson worked odd jobs on construction sites in Nevis. He did block making with Desmond Walters for a while. However, while he was being paid a liveable income, Mr Hendrickson thought it was important that he gained more permanent employment given his obligation to his family, where he had to step in as a main breadwinner in his household for his five siblings. As fate would have it, he was at work and the owner of a well-known construction company came by to look for persons to do an apprenticeship. This encounter with Mr Analdo Chiverton who owned the Chiverton Construction Company, changed his life for the better. He got the opportunity to complete a period of apprenticeship at Chiverton Construction Company and this made a big difference in his career. This is where his training in the wood business started.

> One day I was there and a guy actually came by the name of Analdo Chiverton. Chiverton Construction has been a major part I think in my life – and you know a guy said to me they needed a guy to do apprentice. I looked at it, I was making good money making blocks, I mean the money was really, really good but I left it and I said I wanted to learn a trade so I went along with Chiverton.

He began to learn the trade and gained the reputation of being one of the best roof men in Nevis. Despite the positive response he gained from his customers, his real and most significant feedback came from another professional in the field, an architect. The architect informed him that despite all the plaudits he had received, he was constructing the roofs incorrectly. This feedback did not dampen his spirits but motivated him to better understand his flaws. As such, he had several discussions with the architect, Michael Dore, who gave him a theoretical understanding of the work that he was doing. It was this intervention that had the most significant impact on his career to date. The Island Moldings' boss recounted that this intervention had a telling impact on not only how he approached constructing roofs but also his approach to business in general. He was now able to see the connection between mathematics and angles and its influence on his craft. He further perfected this craft until he was the only person in Nevis who was being contracted to construct the types of roofing which were considered to be the most difficult.

Well at the time there was a lot of style – people don't do much with a lot of style now it's more plain design now, but at that time there was a lot of hills and valleys and different height of roofs. There was a guy who is an architect actually – it's amazing how I learn sometime. I actually took a look to do it, didn't know how to do it, but I believe I could do it and I started doing it, and I was doing it wrong, and there was a guy Michael Dore, he came by one day, he is an architect and he called out to me and say "Hi guys you doing that roof wrong" and I ask him to come and show [me] how to do it.

And actually, it was like a book opened in my head instantly. I didn't know how it happen, I just understood the work instantly, understood angles, he just explained it to me and I just understood.

Working on these construction sites with the professionals and observing how they go about doing their craft, helped to broaden Mr Hendrickson's horizon in the woodwork sector. He now started to look not only at roofs but also learning the art of making doors, windows, and moldings. From his vantage point, he was able to get a first-hand understanding of the industry, and found out that the majority of the moldings, doors, and windows were being imported into St Kitts. He felt that he could produce higher quality furnishings than those that were being imported into the country. So, when there was some stagnation in the roofing area, Donald and some of the other roofers with whom he worked, decided to turn their skills to making furniture. This gave root to his meteoric rise in the business arena today.

In 1999 Mr Hendrickson started out his business under the name, Danny's Woodwork from a small building in his yard. He was intent on gaining a better understanding of the furniture making industry and the latest equipment used in the trade. He revealed that he developed his knowledge by watching videos online, which contributed to his own ideas. Given his strong interest in the industry, it was not uncommon for him to be up at odd hours of the night on the Internet doing research on different patterns and types of machines and equipment that were being used in the field. He even visited the former Premier of Nevis to get some additional support for the business. He was only able to offer encouragement but no financial assistance to enhance the growth of the business. This did not put a stop to his enthusiasm for getting his business up and running. Being the budding entrepreneur he was, Mr Hendrickson used his own financial resources to purchase the equipment he needed to begin the business on a small scale. Subsequently, he sought financial assistance from the Bank of Nevis where he found someone, in the form of a manager who was able to guide him on financial matters.

> I first started at home doing furniture, actually in a small building; and then I remember years ago as a young guy, I used to work on construction sites doing roofs, and I stopped. And this whole idea came about when I looked at the amount of doors that was imported into St Kitts, Nevis.
>
> I actually said to myself these doors can be made locally and that's where the idea actually was born. And, I actually had a meeting with the former Premier, at that time, and I brought up the idea of why not send somebody over there to train with and we can make those we need. They agreed at the time it was a really good idea but I never got any help. So I decided that okay I'm gonna take it up on myself, and I started actually exploring it, going online, looking at machines, looking at videos, looking at YouTube, and a lot of ideas just actually came, and I looked at what machines were used. And I actually used to be up at nights, 1 o'clock to a night from 2 o'clock, looking at the Internet, actually looking at stuff just getting ideas. And, I eventually, you know started to look at the machines and I started buying them piece, piece, you know. And gradually we went to the bank, and it was really good for us at the time, and Mr Royston Isaac was the manager at the time saw what we was doing and he was really interested and he really supported us. So we were able to acquire a lot of the machine that we have.

Moving from Danny's Woodwork to Island Moldings was an easy transition. The name Island Moldings came about because of a conversation with friends about the name of his company. It did not take any major market survey or sophisticated business analyses to derive the now household name for a small company, which has its roots in Nevis but pervades the entire St Kitts and Nevis economy.

> We were doing a lot of moldings. One of my cousins married to a guy name Myrie, and one day he was at work and someone said why not call it Island Moldings.... so he said let's call it Island Moldings...and I just liked it.... Basically, that's what happened.

Product, production, and quality control

His love for all things wood gave birth to a number of products that Island Moldings now boasts as its suite of products on the market. The company makes windows, doors, and moldings all from wood. These are the main product lines now on show.

Over the years because of being on construction sites, I have practically seen some A to Z, in terms of what is needed for construction, and I think that was a plus for me basically in giving me ideas, what I need, things I can do. So what we started was moldings, windows, and doors. We actually started doing that particular wooden type of louvre door, which takes a lot, a lot of skills; and it's not something that you can just look at and get it built. A lot and lot of skills goes into it. My guys here are really, really, really good right now.

Molding is actually used around doors for decorations, they have key areas that crown molding, which goes around your cupboard, your closet and you have the base board that goes around your floor and we actually have about 80 patterns in moldings.... We can take pretty much any pattern and custom design it. If you have a pattern that is not a normal regular pattern, we can custom design anything for you basically, and have it within a week.

Production and quality control

Island Moldings has a very sophisticated production system for producing windows, doors, and moldings. The main input into his production process is treated wood, which he purchases from Guyana or Africa. The African supplies are purchased through a relationship he has with a company in Puerto Rico. The owner remarked enthusiastically that his company does not have a problem with his suppliers as they are very efficient in getting the products on time. He also commented on the efficiency of clearing the raw material at the port in St Kitts. The major concern for him is whether foreign insects and animals such as snakes are present in the shipment, which could deem it ineligible for entry at the ports. This is an important concern given that they are dealing in wood, which comes from the forest and if they are not sorted properly before export, these animals and insects can get lodged into the containers.

> we will be using a lot of wood from Guyana we actually use wood come all the way from Africa, which is really, really good... which is one of the best wood that we use, it's really, really durable it doesn't warp, it's really good.
>
> we have a guy actually in Puerto Rico, and they have a warehouse in Puerto Rico, which actually is based in Puerto Rico but is not really registered as a warehouse so is just a transshipment point actually. So we are able to get the wood at a really good price; it ships straight from Africa, they are the one that actually source it also.... It's not really difficult, honestly. He has a good supply and I think he is really able to get wood sometime, most of the time.

The production process is straightforward. The treated wood reaches the factory and is sorted. The wood is then dressed. After dressing, the wood is placed into the appropriate machine, which cuts the wood to the requirements of the particular product that is being made. After the wood is cut to their appropriate sizes, the assembling of the product is done manually. The owner noted that there are machines that can assemble the product and give a greater level of output. However, the small size of the St Kitts market, which is currently his main market, does not make it feasible to purchase this machine. Currently, they produce about 200 windows and doors per month and 4000 moldings per month. After assembly, the finished product is then sprayed in the appropriate colour based on the requirements of the customer.

Island Moldings has almost all machines that are important for the production of windows, doors, and moldings. Clearly, the owner has spent a significant amount of money to acquire the latest equipment in order to improve the company's efficiency. However, a quick view of the factory showed that the layout of the machines is desperately in need of rearrangement. To this, the owner noted that he engaged a consultant to help with the redesign of the factory space in order to improve the layout so as to meet international standards; and also to create a logical flow from input to output. It should be noted that all the by-products from the production of windows, doors, and moldings are sold to other companies that make various products from the sawdust. Island Moldings does not waste any of its outputs.

> actually we presently supply a lot of the guys, they use it for hot sand in the egg farms or the fowl farm, also for plants. A lot of the products we used is not with chemicals, not full of any chemicals.

In terms of production planning, there is no formal system in place. The company makes doors and windows based on orders they receive. It does not store inventory for doors and windows. However, the components of the moldings, which are less costly to make, are stored in inventory.

Customers and markets

Island Moldings supplies mainly to hardware and department stores (Trade Development Corporation-TDC and Horsford are the main hardware stores in St Kitts and Nevis)[6] and individual retail customers, who place customised orders for the products. Based on his own estimation of the market, Mr Hendrickson thinks that Island Moldings owns about 80 per cent of market share with only two other companies in St Kitts and Nevis in a similar line of business. However, he thinks his competitive advantage arises from the uniqueness of his product, which is a function of the skill level of the company's employees and his adroit focus on investments into his plant. According to the owner, each year, the company takes four to five days for training of employees. This is normally done in the summer period where there is a low level of demand for the company.

> There are other companies that do it [making of moldings, windows, and doors] but the level we have taken it is much bigger ... We have invested a couple million dollars here basically. And with any investment I think that you have to ensure that you get good returns..., with anything you are going to do, you have to have serious investment and I think that's one of the thing that's lacking in the Caribbean ... when I started here I did not want it to be like any other small business and if you look at my building you would realise I did not want to be just an ordinary galvanised shed, which is typical of any workshop that you go into. When you walk into Island Molding you must say, "Wow! This is different!"

In addition to the already large market share that Island Moldings now boasts, it has recently won a major deal to supply one of the country's largest housing developments (the Kittitian Hill Development), with windows, doors, and moldings.[7] This will no doubt increase their market share in a significant way. The owner remarked that Island Moldings was able to win this crucial contract because of the uniqueness of their products and the novel method of negotiation used. Island Moldings was able to present model windows, doors, and moldings to the developers at the time of bidding. This gave them a head start relative to the other competitors for the contract as well as those who were judging the bids had a very good idea of the quality of the products as well as some insight into the ability of the company to deliver.

> your presentation is very important in relation to your product you know, so we try to make sure we have a good presentation also.

Besides the local market, Island Moldings also sells products in neighbouring markets of St Maarten, Anguilla, and the US Virgin Islands. These export markets are not by deliberate strategic actions but came about because of fortuitous circumstances. The owner remarked that while there is no direct strategy for exports, there is still an intention to capitalise on the export incentives from Caricom. There is also the feeling that the push for exports into the US and other North American markets can take a back seat at the moment because there are sufficient markets in Caricom and St Kitts and Nevis to be exploited. The owner observed that Island Moldings has never been out of a job in St Kitts and Nevis since its inception 15 years ago.

> It's amazing because I have 22 guys employed right now and we are never out of work.

As it relates to export development, the owner's thinking on the subject is best captured by what follows:

> Is something that we been looking especially moldings – but I am still exploring it, I am still working on it...One of the thing we look at is, ahm, in CARICOM, Caribbean countries anything that is exported from one country to another is duty free, so we want to capitalize basically on that because rather than bring in moldings and stuff from the US you can actually have it and being locally made export to the Caribbean countries, and all the money stays in the Caribbean, and I think that it is really, really important to keep our money in the Caribbean.

Corporate governance and employee relations

Island Moldings, like most small businesses, does not have formal management and governance structures in place. The owner is the business as is the case with most small businesses especially in the developing world. He is involved in hiring, financial management, production, sales, and marketing, and all operating processes involved in the firm.

> The good thing with me, when I first started out, is that I as the owner/manager, I can do everything in this shop as these guys; and once I bring some guys on board basically I start to feel them out because it's not everybody that has that skill set.

The company secretary and the supervisor in the production area (who has the title of foreman – a traditional description used in the construction industry) are key persons who help in the running of the operation. According to Mr Hendrickson, the secretary and the foreman are key persons in the idea generation process for new products and also novel innovations in the business. He sits with these persons regularly and shares ideas and develops new approaches to doing business.

> I hold the position as foreman of Island Moldings. My job entails that when the boss is not around and even when he is, a lot of the responsibility for coordinating the work – is on my shoulders. I have to ensure that the work is done proficiently, and on time. And to be of assistance to the other guys, whether it be in knowledge or in any other advice they might want.

The employees also seemed very happy with their working environment and the level of leadership displayed by the owner. The foreman especially is a close confidant of the owner and has been with the company from its inception. While he had a good knowledge of woodwork, his main training, according to him, came when he joined Island Moldings 15 years ago.

> upon leaving school I went into the field of woodwork but as a carpenter doing roof work, and then once I joined the company that is where I got my training into molding, joinery, and shop-related kind of woodwork.
>
> I would like to think that I am a very loyal person, and over the years... at Island Moldings he and I have become friends, close friends. And so, firstly, one of the reasons why I stay is because we are friends.

Other employees speak highly about the level of camaraderie among staff and also the friendliness of the owner himself. As one employee said,

> The boss is a good boss and the guys are friendly.

While the company has a good working environment and staff members seemed happy to be in the work place, the company does not have a formal board of directors that helps to oversee its affairs. The small team of the secretary, the foreman, and the owner serves this purpose. Further, the owner usually got advice from his brother, who was actively involved in the affairs of the business until his death just before the writing of this case. With the company growing both in terms of outputs and also employees, the need for proper oversight will become even greater. Mr Hendrickson is aware of this and so the idea of forming a board of directors is not far from his mind.

> Right now, we have my assistant basically and my secretary most of the time and my foreman we would sit down and come up with ideas because what I have realised over the years and I think with any successful business is that you need to listen to your employees... Because sometime they do have really, really good ideas that you never saw all the years you working. When you sit down, sometimes I do have meeting with the guys and I said look, do you have any ideas that we can use and they actually most of the times come up with a lot of good plans.

The company has moved from merely two employees at inception to 22 permanent employees at the time of writing the case. However, there is still a challenge in terms of acquiring the right skills as the company expands. This may even take on greater urgency given the large scale project that needs to be completed at Kittitian Hill.

> I think one of the major thing in any business first thing is getting guys who have the right skill set because that is very, very important. So you must have the right skill set, the guys must be properly trained. Safety is a real issue that can actually shut you down in terms if something goes wrong.

It is clear from the discussions with the owner that he is thinking about something big for Island Moldings as he is not totally satisfied with its current operations.

Future of the business

Island Moldings' owner, Donald Hendrickson is never satisfied with his current position and is always looking at opportunities to improve his current performance. This attitude was very evident in his thinking about the future of Island Moldings. With the tremendous opportunities in the St Kitts market and also the wider Caribbean market, Donald sees the need for Island Moldings to expand its plant, and make business processes and governance more professional in order to take advantage of the opportunities ahead.

> when I look at St Kitts/Nevis there isn't much industries, and I normally look at mine as being one of those that has the ability to export. There is not a lot of industries in Nevis that has the ability but Island Moldings is one of those that has the ability to export stuff, to manufacture stuff and export, so that in any area from tourism anything that's gonna bring money into the country... I think if you have the right plan and the right approach you can approach any bank and get financing.

With the purchase of high end machines like the CMT machine, Island Moldings can also become engaged in the mass production of windows, doors, and moldings. It can also produce furniture and other important household items. The company is looking to leverage this machine to help to diversity the firm's offerings the near future.

> We have the ability to do furniture like this desk, any custom, any type of carving what would normally take a guy to carve a particular design maybe

a week the machine actually do it in 45 minutes. It's really, really good and its very accurate, I mean once you upload the programme you can go and you can repeat that over and over again.

I mean a lot of my ideas actually came from looking at how the Chinese and Japanese, I mean basically the Taiwanese and how they assemble and all of these stuff; and they really used a lot of automated machines in terms of what they doing. But that's something as I said I am actually looking at in the long term.

Business environment and access to financing

It appears that the environment for doing business in Nevis is much more enabling than in most other Caribbean areas. The entrepreneur gives high marks to the Bank of Nevis and other government agencies that have facilitated the doing of business in the country. The acquisition of the expensive CMT machine was financed through the Bank of Nevis. The entrepreneur quipped that the environment for doing business in Nevis was not terribly hostile as his bank was always supportive, the government was not overly bureaucratic, and programmes are being developed nationally to help with training of staff for various industries.

> Basically I would say part of it... doing business... I enjoy St Kitts/Nevis. I mean, I was born here and I like it. It's something that I enjoy and for me one of the thing I believe – my business is not even about making money just dollars alone, it's about – it's based on keeping the money in St Kitts/Nevis, and I keep making reference to that because I think a lot of the times what is done is we send a lot of money overseas but my thing always is to keep your money in Nevis... the Government, I don't have a problem specifically, they support me a lot; they now brought the Prep Programme, which I think is really, really nice where they have guys they gonna train for specific industries and ensure improvements in education all around.

Similarly, the Island Moldings boss thinks that raising financing to support the business will not be awfully difficulty since a number of persons have actually shown interest in the business already.

> Presently, I was talking with some guys who are really interested and there are a couple guys, who are really interested in what we are doing, and they really want to become a part of Island Moldings. We are actually having some meetings presently; I have reached a stage where now I am looking at having some investors I think would be nice for Island Moldings.

Further, while most small firms complain about the rising levels of receivables, Island Molding was not generally pre-occupied with this.

They seem to have found a strategy to overcome the low levels of collection from customers and so the firm does not spend a lot time trying to chase down customers.

> Yearly I think we are doing a couple million dollars, and that's absolutely welcome and that's really, really good. ... We had our share of that problem [collecting receivables] but I think managing, having the right managing system in place is really key. With anything that you have done, or you are doing, there are some key factors that I think is really, really important that makes your customers want to pay you. It's delivering on time, delivering a quality product, 99 per cent of the time they are not going to give you the problem.

Overall, it appears that the challenges that other small manufacturing operations face in the Caribbean are not always present in St Kitts and Nevis. There is no doubt that this environment has stimulated the owner of Island Moldings to think even more seriously about business expansion and growth.

Concluding thoughts

After 15 years of hard-work, overcoming stigma from teachers in school and fellow students, Donald Hendrickson has built a business with EC$15000, which now has a net worth of over EC$7 million. This resulted from dedication, a sense of purpose and initiative, and long hours of work by an entrepreneur and staff that are dedicated to a cause. However, if you ask Donald Hendrickson how he did it, he will tell you one thing, his faith in God. He is a firm believer in the church and has a strong faith in God. He almost surely associates all his success to his belief in God. His final words of advice to Caribbean entrepreneurs are captured nicely in this quotation below:

> I would say the first thing is to get an idea, and I think that is really one of the key thing to any business that you are going to come up with some ideas, believe that you can have it. You can't be afraid of going forward with anything that you going to do because I think that anything that you going to do with your hand ... you must have faith in doing it. Also, anything that you going to do, you got to love it because I love what I do ... I enjoy what I do, and that's real key to any business. ... We have our ups and downs we have some hard times yes but we learn from our mistakes and I think that one of the things you don't keep repeating the same mistakes all the time.

Notes

1. Doing Business 2014 – Economy Profile: St Kitts and Nevis – World Bank, 2013.
2. Compete Caribbean OECS Project – Private Sector Assessment and Donor Matrix Report for St Kitts and Nevis – SALISES, Cave Hill, 2013.
3. Doing Business 2014 – Economy Profile: St Kitts and Nevis – World Bank, 2013.
4. Compete Caribbean OECS Project – Private Sector Assessment and Donor Matrix Report for St Kitts and Nevis – SALISES, Cave Hill, 2013.
5. Ibid.
6. It is important to note that at the time of writing this case, St Kitts and Nevis do not import moldings as Island Moldings supplies all the moldings to the two major hardware stores that resell them to the retail market. Similarly, while there is importation of doors and windows, it is on a much smaller scale. Island Molding estimates it has about 80–90 per cent of that market segment as well.
7. The Kittitian Hill Development will have about 300 housing units. Island Molding will be required to provide windows, doors, and moldings for all of these units.

5
Hot Mama's

Abstract: *Hot Mama's is a small company in Belize which operates in the sauces and spice sub-sector of the tough manufacturing sector in that country. It has become a household name in short order and has seen tremendous success in a tough marketplace. This case traces the early beginnings of the firm, its expansion over time, and provides insights into the future of the enterprise. In addition, it focuses on the firm's products, production processes, supply chain issues, and quality control processes as well. Similarly, the case also looks at the governance of the business and the business environment within which the firm operates and extrapolates how this environment will impact on its future.*

Williams, Densil A. *Competing against Multinationals in Emerging Markets: Case Studies of SMEs in the Manufacturing Sector.* Basingstoke: Palgrave Macmillan, 2015. DOI: 10.1057/9781137500328.0008.

> To be honest with you I really didn't think this whole [business] thing through. I didn't have any business management experience. I didn't have any food processing experience; and to be truthful, when I started out, I could barely cook. I looked at it as something that I would love to do and at first, I sort of winged it most of the time.
>
> Wilana Oldham, Managing Director.

Introduction

The transition from someone who had no idea about cooking, to becoming one of the most celebrated business persons, accomplished by producing and selling items that are solely used in the food business, is indeed a remarkable experience. The story of Hot Mama's from its humble beginnings in a room in a house to its present location on 2.19 acres of property in Floral Park, Cayo District is one of triumph over adversity. There were many hiccups along the way for Wilana Oldham – battling cancer, financial losses, and other setbacks – but she never gave up; today, she is reaping some reasonable rewards from the hard work and effort she put into her company.

Country demographics

Belize is a predominantly English-speaking country located in Central America. Although English is the official language, there is a version of Creole which is spoken as well as Spanish and other languages. The country has a population of over 300,000 citizens according to 2010 estimates, with a larger percentage from rural areas (55%) and the remaining from urban centres (45%). Belize has the lowest population density in Central America, however; it also has a population growth rate of 1.97 per cent per year, which is one of the highest in the Western Hemisphere.[1]

The economy is made up predominantly of exports of crude oil and petroleum, agriculture, agro-based industry, and merchandising. In recent times, tourism and construction have become major contributors to the economy. Agriculture accounts for 11 per cent of the country's GDP.[2] Primary agricultural products in sugar and bananas have been the mainstay of the agricultural sector with the former accounting for half of the country's exports and the latter, being its largest employer. The country has 38 per cent of its lands that are considered to be arable with only 9.7 per cent of the land under cultivation.[3]

As it relates to the business environment, Belize was ranked 152nd in the world in 2012 in terms of starting a business, which represents a five-place drop over the previous year. There was also some regression in the country as it relates to trading across border (107th in 2012 from 106th in 2011) and getting credit (98th in 2012 from 96th in 2011). There are also issues of infrastructure readiness facing the country as there has been a drop in the country's ranking in terms of access to electricity, which fell from 50 in 2011 to 53 in 2012 (World Bank doing business report, 2012).[4]

The hot pepper is a relatively new commercial crop in Belize. There are three major producing areas in the country: Orange Walk, Stann Creek District, and Cayo District. The major market for the hot pepper and its by-products are United States, United Kingdom, and Japan.[5]

The case

Origins and expansion of the business

Wilana Oldham has always wanted to make a contribution to her homeland as she had observed that there was a lot of agricultural produce in the country going to waste and thought that there must be a way to convert the produce into value and make money at the same time. The idea was always in the back of her mind despite the many other opportunities that she got after completing her education. Subsequent to completing high school in Belize, she moved to Louisiana in the United States. During this time she started college but did not complete her programme of study. She worked for a small company in the US as an administrative assistant. Many years later, after she had travelled and lived in several countries around the world, she moved back to Houston, Texas, and that was where she met her husband, Howard Oldham, the owner of a printing company who was also a territory sales agent for Xerox copy machines. Courtship and marriage happened in 1991 and then she joined her husband in his business, assisting him with the running of the office. In 1993, the business copped the number one position in sales in the Southwest United States, and the following year (1994) they were number one in the United States.

Despite doing very well in the business, her husband, an American by birth, had always wanted to live in Belize and with his marriage to Mrs Oldham, the prospect seemed more possible. His consistent prodding for them to move to Belize forced them to think about how they would maintain their lifestyle despite leaving America and its many opportunities. Wilana's mother overheard the conversation and pushed the idea of making habanero pepper jelly, a family recipe that had been passed down. It was at this point that Wilana thought about establishing a business to sell the fresh produce that was in such abundance in Belize but was going to waste most of the time. This idea also resonated with her husband.

The move to Belize took place in December of 1996; they started FOOD Ltd, the original name before Hot Mama's came on stream.[6] The concept of FOOD Ltd was born long before the move to Belize.

> my husband kept talking about wanting to move to Belize and my initial response was "I don't think so!" I was not ready to make the move back... As time went by, he convinced me that we should make the move. Naturally, we needed to figure out how we were going to earn a living. He was not interested in working for somebody else, so it was important for us to establish our own business and make our way on our own. My mother was making pepper jelly at the time; this was a recipe that was passed down through the family from a cousin who came from the Roatan Islands. Just a small amount of cases were made every now and then and were shared with family and friends and eventually she started selling to one of the major grocery stores in the city. She overheard us talking about moving to Belize and got quite excited. "Why not come and make pepper jelly," She asked and I looked at her and very quickly said, "Mom you know I don't cook". A seed was planted and that soon led to further discussions and research as the idea grew.

Coincidentally, Mrs Oldham discovered that Texas A&M University was offering a "Better Process Control" course, which was sanctioned by the US–FDA, and she made the decision to attend to learn more about food safety. It was during this course that she learnt about the habanero peppers, and the new trends of spicy foods from the other participants. It was also established that everybody wanted habanero peppers. At the time, she did not know what a habanero pepper was.[7] The choice of self-employment combined with the newly found knowledge of the new food trends and strong family support hastened Wilana and her husband's decision to move to Belize. Prior to their move, investigations were made into the possibility of the market place. That was how they found their first customer, a food broker in Florida who wanted the fresh habanero peppers. This was the beginning of their business, and for the first few years after actually moving to Belize, they were shipping fresh peppers to the US and sometime later into Canada. The business sold 1/4 million pounds the first year.

This venturing into shipping of fresh produce to North America was the real birthplace of what eventually turned out to be called Hot Mama's. The entrepreneur related her story on the birth of the firm this way:

> Shipping fresh peppers was to be the start of the business venture in preparation for the return to Belize. No one imagined that the fresh pepper business was going to be so hectic and demanding.
>
> The next challenge was to assist with sourcing financial assistance for the farmers so that they would have the necessary means to maintain their fields throughout the project. B.E.S.T. (Belize Enterprise for Sustainable Technology), a non-governmental financial entity that was in the market to assist minorities, etc., became a part of the project and they provided short term loans so that the farmers could invest into growing peppers.
>
> Mash became the second product. A Hobart meat grinder was purchased and those peppers which were not exportable were ground into pepper mash. A new agreement had to be made with the farmers informing them of a change in the receiving of the peppers, but which was necessary to save what produce was brought in.

Belize by now had developed a reputation for producing the best habanero peppers. As the entrepreneur related, whispers had it that some of the buyers would not purchase peppers unless they came from Belize. Business seemed to have picked up well for the entrepreneur but all was not as positive, as conditions changed over time and things became more difficult.

> The business of shipping fresh peppers was good, with regular weekly shipments being made not only to the USA, but then also to Canada. Despite all of my efforts in building the fresh business, no loyalty was given and eventually I lost out to other individuals who promised higher prices for the supply of fresh peppers. Changes were coming about and eventually my shipments became less frequent with much smaller amounts. The decision was eventually taken to get out of the fresh shipments.

Not one to be daunted by less than favourable business conditions, Wilana turned her attention to how best to diversify her business to ensure its survival. This is how she related the reactions to the changes in market conditions:

> Additional products that came from the making of pepper mash are the dried peppers and pepper powder and these turned into another segment of the

business. The introduction into this area was due to a visit from an engineer out of Guatemala, who invited us to make a trip to Petén due to an issue they had and they wanted and needed my advice. What they found was 200 farmers growing habanero peppers, but who did not have a market. They were shown a room that literally had a mountain of fresh peppers sitting on the floor. They wanted to ship fresh, but did not know how to. It was also found out that they could not ship fresh to the USA, because they did not have the clearance from US-FDA for the area. Due to the urgency of the matter (peppers already sitting on the floor and much more to come in) it was suggested that they immediately grind the existing harvest into pepper mash. A grinder was found, drums were bought, a formula was shared and the making of pepper mash was on its way. What to do with the new inventory of mash? I made some phone calls and eventually spoke with individuals from the McIlhenny Company (Tabasco), who promised to come to Guatemala to visit the fields. Everyone was excited about the prospect. McIlhenny could be the solution to all of the problems. Unfortunately, this lead did not pan out as (Tabasco) was in the early stages of product development and the market place was still needed for the consumption of habanero peppers. Luckily, I had also made contact with another individual from Georgia, who sold pepper mash all over the world and started negotiations with him to become a supplier of mash. He was happy to work along with her and also made the trip down to Belize and over to Guatemala to look at the fields and discuss the potential. Agreements were made and signed and new fields were put in the ground.

through Internet search I found a company up in northern part of the USA that did refining and extraction. They became very interested in the pepper powder and samples were sent for testing. Eventually, they purchased all of the habanero pepper powder that had been produced (20,000 pounds).

The business had now fully moved from the shipping of fresh habaneros and more into the higher value products of powder and mash still made from peppers. Similarly, the supply had shifted over to Guatemala. As the business expanded and became known, other customers started sending orders for other products such as pineapples as well.

However, as with most aspects in the entrepreneurial journey, there will also be ebbs and flows. Overcoming challenges had now become a part of Wilana's modus operandi. As her business started to recover from the first shock and again moving into its new phase of operation, a personal tragedy struck. Relating this story was not easy for the entrepreneur.

> tragedy struck in the form of cancer...quick decisions had to be made and I left for the US to seek out medical advice and eventually treatment. I

came home after six months, weak but determined to get back to my life. Everything was still on hold as it was necessary to travel back to the US every three months for checkups. It was between one of these three monthly visits that a two hour trip tubing down the river ended up being six hours and I was severely sunburnt. Three months later, a noticeably spot was noted on my leg, but it was not for another three months later before any attention was given to it. I had my second cancer in the form of melanoma. This led to immediate surgery. A year later it came back and once again immediate surgery followed by a year of chemotherapy. It was a further six months before I could say that I was ready to be back to work.

Despite the challenges with cancer and the almost total loss of her business, the strong and determined entrepreneur did not decide to give up. She was determined to continue to make a difference with her business. This is how she recounted the story:

It was June and mangoes were in season. Anxious to get busy again, Howard suggested that I should start working on a recipe for the mangoes...after all, they had the name but no product. In my home kitchen, I started developing the Manganero sauce and through many trials and errors, eventually settled on something that I liked. Samples were given out and great responses were received, but the product was not stable enough for store shelves. It took time, but the desire to produce something that did not require chemical preservatives, but was still shelf stable was upper most in my mind...For the next 12 years, I continued with frequent trips back to the US for checkups and treatments, but my passion to grow the business was stronger than before. It was right around the year 2002, having moved completely from the fresh produce to then producing mash and finished products that the name change took place from FOOD Ltd. to Hot Mama's. It was through meetings with friends and business associates from my husband's social, community based club, the Rotary Club of San Ignacio, that the idea of a name change was born. It was after an exchange at one social meeting that the name change went from FOOD Ltd to Hot Mama's.

Business growth and expansion

When the company changed its name in 2002, it was time for it to move into the next level of operations, which was expanding the range of finished goods. Already included in the list were the pepper mash, the whole dried powder, and the pepper jelly. This period brought strong benefits to the business. Moving from just two employees in early 2002 to nine fulltime employees in 2014 is not a bad result for a small business. However, it seems as if business growth was happening serendipitously without much deliberate thought being put into expansion.

There has been no written business plan and decisions were being made as situations arose. Many times a wrong direction was taken and the need to start over was apparent. Do I wish that I had done things differently... of course... but it is not possible to change that at this junction. All that can be done is to accept, learn, and move on from it. With much assistance and guidance, more structure is being put in place to ensure a more smoothly run business.... Moving gradually from just two people to a team of nine is a great improvement... the next step is to increase ever so slightly the team, but most importantly to improve the working conditions and capabilities to improve the working situation and outputs for the company.

Products, production process, and quality control

Today, Hot Mama's Belize is into the business of selling finished products using traditional Belizean fresh produce as their major inputs. It sells mango sauces and pepper sauces, among a range of other items. Besides sauces and spices, the company has also diversified into other areas such as farm tours, a clothing line with the Hot Mama's label, and confectionaries for visitors to Belize.

> I was making pepper jelly and it was the first time I did a double batch. I packaged it and everything, and I came back later on to check on them. I did this and I was like "uh oh, it is still running." So I went crying to my husband, "oh, Howard I can't sleep, I just did a double batch – I know you are the expert in charge, it's not gelling, and I don't want to throw it way. I would hate to throw it away, it is so much money." You know what... he stood there listening to me for a while, and he said, "Are you finished?" I said, "Yes, I don't know what I am going to do!" And then he said, "well, why don't you put it in a ten ounce bottle and we call it sweet pepper sauce." And I was like, "What! No we can't do that!" Of course, I ran to my mother, with the suggestion and she said... "Oh no, no, you can't do that, no, no." But my husband made a suggestion and so I said okay, we'll do it. And the sweet pepper sauce was a success and is extremely profitable. It is our number one sauce today.
>
> not everyone wants to eat something spicy for breakfast and that is how we ventured into the seasonal/tropical fruit jellies. We wanted to capture more of the breakfast market with the understanding the foreigners coming to Belize wanted the entire Belizean experience and that included eating our delicious

and exotic fruits. Naturally, what followed next is the dark chocolate with the habanero peppers. Yes, we like to spice things up a bit for our customers!!! .... my inspiration comes from within...hoping that my tastes will be liked by others as well.

Production and quality control

When the peppers are brought in by the farmers, a thorough inspection is done to ensure that they meet the required standards before they are processed. The peppers are washed, sanitised, grounded, and blended with salt and vinegar to preserve them. To maintain traceability of the supplier, each farmer's delivered peppers are kept separated and a recording log is kept with the drum as it is moved.

There was also the issue of logistics to get the produce to Hot Mama's; especially since the price of fuel is high and some of the farmers were located in distant, isolated areas. At times, it was necessary to reject the produce, especially during the heavy rainy seasons, but with some patience and constant communication, the farmers eventually learnt the importance of proper post harvesting; and today the situation is relatively worry free.

> There was a time when I had to do a lot of rejecting because peppers were coming in wet. The peppers were breaking down as they were coming in. Peppers like a lot of water, but they don't like sitting in water and they don't like a lot of moisture on them after they have been harvested. To avoid misunderstandings with what is to be delivered is an important step in the relationship with the supplier.

The relationship between suppliers (Belizean farmers) and Hot Mama's Belize can be described as excellent, thanks to the continuous support and guidance Mrs Oldham provides to them for the continued improvement in their yield. Importantly, she also attributes this to her paying her farmers on time, and they have a tremendous amount of respect for this practice.

The production process at Hot Mama's Belize is indeed very labour intensive. However, there have been efforts at trying to improve this. The company received project funding assistance through Caribbean Export in order to mechanise their bottling operations. A filling machine was purchased from abroad, however, there was an oversight in the project as it did not specifically include the services for the technical expertise to set up and train the end user on operating the machine; therefore, that component was not funded. It was felt that sufficient due diligence was done, because the team had visited the manufacturer's plant to get a first-hand orientation of how the machine should operate, and was told that it was a simple procedure. The machine was bought and shipped; however, upon receipt of the machine, several issues developed and caused it to malfunction. Additionally, it was discovered that the manual was not complete and is considered to be more of a technical manual than for the operation of the machine. Efforts were made to contact the supplier of the machine who had made promises that had not been fulfilled, which included sending a complete manual with a 10+ step guide to operating the machine that was still not forthcoming. A local electrical/mechanical engineer was then hired to assist with the equipment and who was able to get the equipment working, but only for short periods of time before it started malfunctioning again. Today, the team is negotiating with the manufacturer to bring a qualified, experienced technician to do several things:

1. Check the set-up of the equipment.
2. Train one or more employees on how to run and use the equipment.
3. Facilitate the production of a video which can be used in the future for training, but also to aid in addressing any issues that may arise.

The Hot Mama's Belize team has in recent times implemented the procedures and requirements needed to adhere to the principals of HACCP (Hazard Analysis Critical Control Procedures). Certification will be their final goal, but in the meantime, the manuals have been written and are being adhered to. The company has taken a proactive approach to ensuring traceability of all their products from "farm to fork". In an effort to achieve this, Belize has prioritised the training of farmers in new farming techniques as well as different ways of producing peppers. Historically, the farmers often chose to grow peppers in the rainy season due to the absence of proper irrigation systems. The training has been aimed at improving the knowledge of the farmers in the planting, harvesting, and selling of their produce. Critically, for traceability purposes, the firm has to ensure that the peppers from each farm are not mixed as well as to ensure that if there are any harmful chemicals existing, that it can be traced to the source. Wilana's reflection on this is captured as follows:

> it is very important to understand what has happened in the field and the push for compliance cannot be stressed enough. The records from each of

the fields help with the traceability and when more than one farmer comes in with their peppers, each is accepted, processed, and recorded separately. That record is kept with the pepper mash until it is used in the sauce.

Although not yet HACCP certified, the team at Hot Mama's Belize realises the importance of this process and have made every effort to adhere to its principals. They have successfully implemented the GMPs (Good Manufacturing Practices) and SSOPs (Security and Sanitary Operating Procedures), which are the first two components for HACCP certification. The team started out by gaining the 5S certification which is the first step towards certification. This step deals specifically with sanitation and layout; that is, keeping the facility clean, organised, and clutter free. This is an ongoing project and it ensures that any other potential procedures will be easier to adhere to.

Market expansion and customers

Hot Mama's is mainly focused on selling into the Belizean market at the moment. Its products can be found in most of the major supermarkets, grocery stores, mom and pop shops, airport, and hotel gift shops across Belize. The products are presently being distributed by the company through the means of a salesman going door to door (stores). In the past, the use of a distribution system was utilised, but over time, it was felt that better success could be attained by the company moving the product itself. Even today, despite marketing and promotion, there is still a bit of confusion by some consumers thinking that the Hot Mama's products are creations of its main competitor, Marie Sharp.

> When the decision was made to move into the finished products area, it never crossed our minds to look at what the competitor was doing or what products she was producing. Yes, we knew that she was producing pepper sauce, but it was felt that the products were to be made would be different. As time passes, it became gradually clear that there were two types of taste buds on the market... those that liked it sweet and those that liked it salty. At first, the lean was more towards sweet and hot products... but it was not too long afterwards that the push was for hot and salty items too. The gourmet sauces which are the most successful sauces... more successful because of their uniqueness and versatility in use and the consumers in particular ask for them when they enter the stores. The idea at first was never to produce

the pepper sauce and so approaching the competitor to private label her sauce with my brand was the thought... after all she had been producing habanero pepper sauce for thirty years plus – a well-established sauce. At first it worked, but it was soon discovered that there was a feeling of unhappiness on her part and eventually the move to produce our own sauce was necessary. In the development of the pepper sauce, every effort was made to ensure that the product was different... less salt, less vinegar, more peppers. Despite all of that, people still think and ask if the product is made by them. Today, the company focuses more on making the distinction between the two companies by promoting its gourmet sauces. It has recently added a line of seasonal and tropical fruit jellies... Once again, ensuring that the varieties are different from the competitors.

It was important for Hot Mama's to capture a sizable portion of the Belizean market place for its sauces. Every effort has been made towards direct marketing of its products by going door to door. The result is an increase in sales since this new shift in strategy to direct marketing. Further, as part of its efforts to gain customers, the company also got involved in a number of corporate social programmes as well. These include sponsorship of club and societies events, football and sporting matches, school promotions, fairs and trade shows in the community among other things.

> at first it was mostly about sponsoring sporting events, but eventually we moved into radio ads, a lot of giveaways, especially to entities going aboard to trade shows that promoted Belize.
>
> the Belize market is very small and limited, and in order for the company to grow, it is necessary to consider exporting. Not having a proper export plan in place, which can be followed, can quickly lead to chaos. Understanding the limitations and constraints has pushed for new goals and desires and quite possibly re-inventing itself. The name of the company is well established now, but the use of the name outside of Belize can be limiting; therefore it was realised very quickly the need for a well thought out export plan is critical. The first six years of operation was more about establishing the name, the brand, and the products.

Regarding the competition, there is Marie Sharp who has a long and established history in Belize. There will always be new players into the market, but due to the change and direction of Hot Mama's, it is not felt that any major impact will be likely.

> in Belize there is Lisette's Secret Sauce, which is a division of Bowen and Bowen. The company is a well-established company with many divisions to

include carbonated sodas, water, beer, ocean freight, vehicle sales, and services. They do coffee and a couple sauces but are not pushing those items too hard. GraceKennedy (Belize) a distribution company and who has sauces that are brought in from Jamaica. Occasionally there are Matouk's coming out of Trinidad, but quite possibly the cost of logistics may make it cost prohibitive for other sauce companies entering the market place.

Despite the various competitors in the market place, the company is not taking market expansion and penetration lightly, and is constantly reviewing and revising its plans to improve its market share despite its present relatively good position in the market place.

> Today is whole different story. A year or so ago, there was a sense of unease. Where was the company going? How was it going to grow to achieve the vision and mission statements laid out a few years back? There was a sense of frustration as the company appeared to be stagnating. So much effort had been put into brand awareness but not as much effort had been put into the distribution and market penetration. With the purchase of a new cargo van, the hiring of a salesman and the concept of direct marketing, it has led to increased sales through aggressive efforts which is now paying off. Now the consumers, both local and foreign are commenting "there is another company out there that makes unique and delicious products".

Interestingly, Hot Mama's has also become more aggressive in attacking its main rival, Marie Sharp, through a commercial directly aimed at them. This is a bold development for the company, which has always taken a more laid back and unaggressive attitude.

> the focus is now to get onto the tables in the restaurants. We want people to be eating the sauce and not just seeing it on shelves and so a special offer is made directly to the restaurants enticing them to start putting the sauces on the table. It is a little bit of a challenge since they already have Marie Sharp. No one is asking you to get rid of Marie Sharp; just asking you to give your customers a choice by putting the Hot Mama's sauce on the table.

Export markets

Besides the local market, Hot Mama's has also done a fairly small amount of exporting. Given its strong focus on quality it has always had a competitive advantage to sell its products in the export market. Its products are marketed mostly in both the United States and certain areas in Canada. While the presence in the US market is still not high, there is still an interest there.

besides the local market, small shipments into the US and Canada are being achieved by using the services of the post office. Container quantities of pepper mash shipments have been made to Barbados. A newly formed partnership with a group in Petén, Guatemala, promises to open a much larger market to the Hot Mama's products. This was realised due to the proximity of the facility on the main highway that runs between the countries of Guatemala, Belize and for access to the Yucatan in Mexico. Most definitely the move to penetrate the Guatemalan market is being assisted by the flow of traffic by the Guatemalans as they travel by the facility on the way to their destination. Word of mouth has been a big plus and the question is always when will the products be available in their country?

I don't have a huge presence in the US and that's a challenge. It's a huge market and there are a lot of trials in going into that market place, especially going into a grocery store, which is a whole dynamic by itself.

The real opportunities for expansion into the US are for the company to offer its products to others by private labelling. The cost of logistics, marketing, distribution, and any liabilities then becomes the responsibility of the newly branded products.

> Private labelling is a win, win for Hot Mama's. Increased production and reduced expenditure means increased cash flow and happy working employees.

Corporate governance and employee relations

Similar to most small businesses, all of Hot Mama's decision making on the operations of the company rest with the owner. While Mrs Oldham consults from time to time with staff and other persons about business decisions, the ultimate decision has to be determined by her. Clearly, this structure places enormous pressure on the owner to think strategically while dealing with daily operational issues.

> I have failed miserably as an administrator, but please allow me to bring some clarity to that statement. The management of the company has always been held closely by the owner. Tremendous time and efforts has been put into the mechanics and it is simply difficult to let it go. Understanding these shortcomings and needing to make the company grow, change must and is happening. This has led to the hiring of a production manager and eventually the Director of First Impressions.
>
> I am now making a shift with Michelle by teaching her how to manage the day to day running of the office. She already knew quite a bit, but now the

shift is allowing me to focus on some of the important areas of the business, such as the construction, writing proposals and meeting with clients.

The company does not have a Board of Directors nor does it have a formal organisational structure. Mrs Oldham remarked that her Board meetings take place at evening times with her husband as they discuss the day's events. This aspect of the business is also changing as she has come to realise that there are benefits in having a Board; meaning that she does not have to be the only one to come up with ideas and suggestions. Having input from others also means that there are always new thoughts and a different perspective on the situation.

> My husband is my number one fan, and has been very supportive, not only financially but also emotionally. Even though he is not involved in the day to day running of the business, we always discuss the on goings in the evening time. We would laugh and joke about having a Board meeting at night. He has been most instrumental in assisting with the infrastructure, he is the main contractor and whenever there is a problem he helps finds a solution.

The staff enjoys a strong level of camaraderie, which helps with the running of the organisation. They often bond together to deal with issues and challenges and in most cases, they will find a solution to the problems encountered. This spirit of "oneness" has helped Hot Mama's operate more effectively without a formal organisational structure. While there is no formal organisational structure, there are some interesting titles given to members of staff. One of the most interesting is "Director of First Impressions". This person is responsible for the gift shop, the online store and tours of the property. This position seems to be key to the organisation. The story behind the hiring of this director, Michelle, is quite remarkable.

> I was interviewing her for a position in the factory and she was very upfront and honest with me about what she was capable of doing. But in my mind I thought, I have got to hire this girl one way or the other. But it was a month later before actually being able to create a position to her. She has been with the company for over two years now and she is my rock... When discussing Hot Mama's, it is more about the team, because they really and truly make the company what it is today.

While the staff enjoys a close bond, the hiring practices at Hot Mama's are also not very structured. The humanitarian personality of the owner generally overrides a strategic approach to the hiring of staff.

I am quite a softy when it comes to hiring. Very rarely do I stop and ask "are you qualified, can you do the job"? Instead it is mostly saying "No, no, I will train you." The desire to help everyone is a real issue since their lack of experience becomes a cause for irritation later.

Future of the business

While Hot Mama's has done a remarkable job to move its operation from a small trading company, which sells fresh produce to becoming a well-established household name in the sauces and spice sub-sector of the manufacturing sector in Belize; there are plans to move beyond its current operations and to capture greater market share both locally and internationally. This entrepreneur is not one who remains comfortable with her present position, and as such, she is always seeking ways to enhance her business growth.

A partnership has been formed in Guatemala and the concept is to eventually produce the sauces there. Initially though, everything will be made in Belize and shipped over. This will allow the new company in Guatemala to develop its distribution routes and build up its sales, but at the same time, it will give Hot Mama's the benefits of the sales and the production. Guatemala is an exciting market, simply because of the population size. Eating spicy foods is also something fairly new, but the demand is growing. An introduction has already been made with one of the gourmet sauces and every day calls are received asking for product availability. Presently, the focus is on completing the legal aspects (documents, licenses and permits) to ensure trouble free border crossing.

Recently, the company has started discussions and negotiations with a medium to large salsa company in Texas. They are most interested in securing a supply line for habanero pepper sauces since their plans are to expand their markets. A visit from leading key personnel has move the interest to one of the gourmet sauces and that has promoted a request for that product to be included in the shipments. Only a 40 foot size container can be shipped to them, and they want the products in bulk. Bulk packaging will be new to the company, but it will mean that it will be necessary to increase the capacity levels to meet this new request.

The business environment and the future

While there is a bright future for Hot Mama's, the future could have been even brighter had there been a more enabling environment for businesses in the domestic economy. Clearly, the challenges of doing business in Belize have slowed the growth of the firm in a significant way.

> Learning to be self-sufficient was an important step in making things happens. When anyone becomes too dependent on others or the government, the risks are higher for disappointment. ... The domestic banking system has not helped much as well. It has limited the growth potential of the firm. The cost of borrowing from the commercial banks is extremely high. In addition, they only loan for short periods of time; thereby making it very difficult to manage the finances. The need to investigate other potential sources of low interest loans or funding is necessary.

Mrs Oldham is not a big fan of commercial loans, but sometimes feels that they are necessary to allow the company to grow. That is the situation she is in presently. To penetrate the external markets, certain expansions will be needed and that will require an injection of new funds to make it happen. This will require some careful planning and a dynamic proposal.

Concluding thoughts

Hot Mama's is a story of triumph over adversity. Being diagnosed with cancer twice, it would be easy for any ordinary person to give up and live a quiet life. However, Wilana Oldham is not any ordinary person, she is an extraordinary entrepreneur. She has decided to make a contribution to her homeland Belize, and nothing has stopped her, as she launched and grew her business from selling fresh produce to becoming a household name in the sauces and spice business locally and internationally. Despite the tremendous success to date, she is one who is never comfortable with the status quo and is still in search of bigger opportunities for business.

Notes

1. Information from the 2010 Census, Statistical Institute of Belize.
2. Central Bank of Belize.
3. Information from Beltraide.
4. "Doing Business in Belize: 2012 Country Commercial Guide for US Companies" published by the US Department of State 2012.

5 "The Hot Pepper Industry in CARICOM: Competitiveness and Industry Development Strategies" by Singh, Seepersad, and Rankine 2007.
6 FOOD stands for Fuller Oldham Overseas Development. It is the surname of both Wilana and her husband.
7 Habanero is commonly referred to as hot pepper. In Jamaica it is referred to as the scotch bonnet pepper. The scotch bonnet is a part of the habanero family.

6
Perishables Jamaica Ltd

Abstract: *Perishables Jamaica Ltd is a small firm which manufactures and sells tea locally and internationally from its home base in Jamaica. It is one of many small manufacturers in the difficult manufacturing climate in Jamaica which is not only surviving but prospering against all odds. The case highlights the early beginnings of the firm, its expansion over time, and provides insights into the future of the enterprise. In addition, it focuses on the firm's products, production processes, supply chain issues, and quality control processes as well. Similarly, the case also looks at the governance of the business and the business environment within which the firm operates and extrapolates how this environment will impact on its future.*

Williams, Densil A. *Competing against Multinationals in Emerging Markets: Case Studies of SMEs in the Manufacturing Sector.* Basingstoke: Palgrave Macmillan, 2015. DOI: 10.1057/9781137500328.0009.

> One, one coco full basket....we as a company when we started out in 1980 we started our company with JA $300, which was the equivalent of US $170.... Today 2014 we are doing J $80 million in sales annually.
>
> Norman Wright, Managing Director.

Introduction

Perishables Jamaica Ltd, 2 Leonard Road, Kingston 10, a small manufacturing company in the hot beverage business represents a story of struggle, perseverance, hard work, dedication, and commitment to one's dreams. Even with ill-health, the determination to fulfil one's dreams never dies.

> as I am speaking to you, I am student of UTECH[1] and I am completing my Master of Science in Complementary and Alternative Medicine but for illness which came along in recent months I would have completed my final semester.

Norman Wright's family background, his community upbringing, and his educational achievements are all factors that have led him to venture and start his own business. His background has also contributed to his dedication to move the business from a mere start up to become an established enterprise competing with larger players in the marketplace. From its humble beginnings 34 year ago with US$170, the company has grown to become a household name among the large and more established operations in the Jamaica and Caribbean landscape that are engaged in the tea business. It also has a burgeoning export business which supplies customers in North America, Europe, and the wider Caribbean region. This is a remarkable story for a company that started out with a single vision, to use local herbs to make an economic contribution to the development of Jamaica. This did not happen purely by chance; rather, adroit leadership and strategic thinking from the principal owner and his team led to this impressive performance over the years.

Tea industry in Jamaica

Global value of the tea business has been calculated at between US$30–34.5 billion, between 2010 and 2013, with the projection slated to reach US$37 billion by 2015. Locally, the yearly imports of the commodity averaged approximately US$1.5 million with the US$7.7 million of imports recorded over the period 2008–2012. A big part of the development of the industry is the import substitution trends as currently some established international brands such as Bigalow, Celestial Seasoning, Twining, Uncle Lees, Waitrose Love Life, Arizona, Cozy, and Asia are a part of the local market. However, there have been local brands which have entered the fold and acquired a share of the market, with the emphasis being placed on the export markets regionally, in North America, and in Europe. Some of these major local brands include Caribbean Dreams, Tops Teas, Salada, Kendal, GraceKennedy, Good and Natural and, Sipacupa.[2]

The case

Business origin and expansion

Perishables Jamaica Limited started in 1980 out of an opportunity which the owner and entrepreneur Norman Wright spotted during a period that could be described as an economic crisis in the Jamaican economy. Having armed himself with graduate qualifications in both the natural and social sciences and with strong work experiences from established enterprises, the visionary entrepreneur decided that out of every crisis there is an opportunity and so, he decided to start his own company in order to take advantage of the benefits the crisis in the foreign exchange market in Jamaica offered. The narratives below provide insightful information on the business origin and growth.

> I went to York Castle High School in St Ann and I left there in 1971 and went to CAST[3]... and I spent a couple years there and did a lab technology course I left from there and went to Esso oil refinery as it was then... Which is now the Petrojam refinery down at Marcus Garvey Drive, and I worked in the lab doing research and bench work for a couple a years, in 1975 I went to work in St Thomas with Tetley Tea and Orion Sales doing quality control work; so my empirical base is in quality control.
>
> I subsequently cross trained into industrial management at CAST; I cross trained into Accounting at CAST; I cross trained into Marketing at CAST, so I built up my educational base in areas of weakness that I had in my own personal qualifications: And that allow me the opportunity to progress and move through the organisation in Orion Sales, to move from

the chief chemist to become the production manager, to become the factory manager, to eventually being the general manager and a director of Orion Sales; and sort of simultaneously I was doing the same thing because Tetley Tea Jamaica Company Limited and Orion Sales were sister companies in many respects sharing similar shareholders and management up to the point in time when the Tetley Tea Jamaica Company Limited was divested to T. Geddes-Grant.

Interestingly, although doing very well in his employed positions at the various companies for which he worked, Mr Wright was not settled and had the latent entrepreneurial spirit looking for the correct time for it to be unleashed. The foreign exchange crisis in Jamaica in the 1980s was a turning point for that entrepreneurial spirit to become alive. While the external environment was a fillip to getting Mr Wright to start his own business, similarly, other factors such as his family background helped to determine the line of business that he pursued.

> I had a close connection with farming. My father worked with the Ministry of Agriculture at different times and the Ministry of Industry and Commerce as a Produce Inspector and as an Agricultural Inspector and Agriculture Officer for the Ministry of Agriculture, so I was involved in farming. My grandfather was a farmer on one side, that's my dad's dad was a farmer, I came from those humble backgrounds.... like leaving from home in the mornings with no food in the house and walking to school. We didn't have the privilege of taking buses like we do now.

The literature is generally silent on the role that an entrepreneur's faith in God plays in guiding how some owners are able to deal with the challenges of running their own operations and grow their small firms to become established enterprises. Nevertheless, like other entrepreneurs, a strong faith in God has played a role in how Mr Wright views his business dealings and how he relates to business and professional associates as he grapples with the challenges of running his small firm and surviving in the highly competitive tea sector.

> mother's father was a Methodist Minister and so I do have my own religious convictions and my time at York Castle High School was spent living with my adopted grandmother/grandaunt who was married to a Baptist minister. So you know, you were exposed again to the religious convictions.... We are very upstanding and fair to who we deal with even if they treat us unfairly, we don't really worry too much about that; we just make sure that you maintain your goodwill and that is really what keeps us strong as a company and keeps us honourable as a management.

In addition to upbringing and family background, one cannot underestimate the role of previous work experience in helping small firm owners to establish their own businesses which eventually survive and prosper in highly competitive market spaces. The work environment and the core values of an individual do help to shape the type of business they enter into. This was evident in the case of Mr Wright. He did not merely start his own business out of a love for money, although that was important, but there was also the strong need to make a contribution to the community and country in which he grew up. The narratives below provide a rich account of how previous work experiences have helped Mr Wright to establish Perishables Jamaica as a sustainable business and not just a hobby business.

> I moved from in Orion Sales/Quality Sales. It was Quality Sales first then Orion Sales. They changed the name and simultaneously working with Tetley Tea. I moved from quality control into production management, into plant management into general management, and I became a Director of Orion Sales in about 1980. I was a Director for quite a few years up until about eight years ago I gave up the position. I, at the same time, did consultancy, management consultancy work for Kiwi Caribbean, which was a subsidiary of Sarah Lee/Doug Egberts, and I did that for about nine years working with Kiwi Caribbean. Simultaneously, I was Director of Orion Sales, at that time I gave up some contact with direct management with Tetley Tea but then while I gave up direct management to them, they were contract packaging products for me, because I had at this time developed Tops Pep-O-Mint,[4] which was using indigenous raw materials. I saw the need to try and get Tetley Tea to integrate backwards to use domestic raw materials to create a Jamaican product and that we did successfully, and that product is still being made and exported. That pretty much – constitutes my sojourn into work.... I had been running Perishables Jamaica Limited as a hobby business for many years. And, I decided that it was prudent for me to go into it full time, and it offered me more scope and personal development in terms of creating products that were using Jamaican raw materials and we weren't importing raw materials to produce products that we could produce locally.
>
> After a while it get to you, so you are not doing anything to help the people, you only making a money off them and that not really, that's not my personal objective in life now. I pray for health and strength and daily food I never pray for wealth and strength yet. So, I figured it was time I move forward and do something that would have a significant and longer lasting impact on a population, so I went full time into herbal tea business and created products that use raw materials from Jamaica and export them... and we work with small farmers in much the same way that I saw my dad working with small

farmers over the years because I know that they have a tremendous contribution to make to the Jamaican economy.

The growth and expansion of the business was not an effortless undertaking for Perishables Jamaica. The owner's leadership was critical in moving the business forward. He was always vigilant in looking at the situation in the domestic economy and the developments in the macro-economy and then used that data to support his decision making in his firm. However, what was at the forefront of his mind is the need to continue to develop backward linkages to ensure that not only Perishables Jamaica and its owner(s) benefit from being in the business but that the communities in which he operates and the wider economy benefitted as well. The budding entrepreneur used his knowledge gained from his previous job experiences to help to inform strategic decisions he made to grow his business operations.

> as a young manager at Tetley Tea and Orion Sales at the time, I saw people coming to work without any work to do, and we had to send them home and lay them off because there was a lack of foreign exchange... as a company, as an individual, I saw the need as I said, to look at domestic capital formation and so we set out to create the products that would use local raw materials and which would provide us with some backward integration of our operations so that the farmers would be involved and that we could create value added products from indigenous raw materials that would not only be sold locally but could be exported.... And, that is what we have done continuously over the years as a company.... we have moved from 1 employee to about 16 employees now, and we have moved from dealing with 4 and 5 and 6 farmers to dealing with probably, at our peak we were dealing with 800 farmers.

Despite all of this, the real moment for the branching out into the tea business came when the owner had serious problems in getting his export business to be competitive in order to survive in the market place. Perishables Jamaica did not start out as a tea business but was mostly into the business of exporting Jamaica fruits to North America. How the name perishable came about is not a mystery. It evolved from the company's exact line of business that it started in 1980.

> when we developed the Pep-o-Mint product we originally were getting into what was fresh food export; that's how we started out within 1980 – hence the name Perishables Jamaica Limited.
> we packaged our mangoes, and naseberries and our avocado pears and that sort of thing and send them overseas. Whenever we shipped out anything we were always getting the feedback that of the 20 cases or 50 cases of products

we sent up only 60% of it was good, the other 40% wasn't good. And as we went through that experience in the first couple of months I made a visit to Canada once – I made sure that I went on the plane that the stuff was going on. So I dropped it off at the airport, I took the plane up and I went to the airport to see the thing being cleared the following day, and what I saw in Canada and what I shipped in Jamaica I realised that it wouldn't be prudent for me to stay in business because the boxes were all smashed up and the product was all battered and bruised because of the handling that took place even though we had make effort to package properly in Jamaica…. So I realise that wasn't the way to go. In speaking with one of the storeowners they suggested to me that we should since I was at the time working with Tetley Tea and I was involved in the tea business that I should look at developing a Jamaican Peppermint tea bag. And Senator Anthony Johnson at the time when we were having problems getting foreign exchange to buy black tea and we had to be working 3 days a week at the Tetley Tea plant or 2 days a week, he said you should look at developing orange tea and those teas. So with my experience in farming, my experience in quality control and manufacturing, and whatever experience I had I decided to work with my dad to source some peppermint and dried it and got it grounded and started to package herbal teas, which are dried products which have their own problems of E.coli, and salmonella and how you dry and process and handle them but you are less likely to lose as much as you would with the fresh produce.

Products, production, and quality control

Perishables Jamaica could be described as an agro-manufacturing company. It draws its main inputs to produce its teas from the agricultural sector and then manufactures these into high value added products which it sells to its consumers. Using the jargon of value chain analysis, the firm could be seen as a value added company to the agricultural sector. The company has a list of products which are generated from Jamaican herbs and spices. The narrative below provides a good overview of the wide portfolio of products that are on offer.

> It's basically, essentially herbal teas that we are focusing on; ginger and its by-products, ginger with peppermint, ginger with lemongrass, ginger with cinnamon, ginger with sorrel. Then we have our peppermint products; Pep-o-Mint, which is a blend of black tea and Jamaican peppermint. That was our first product, and when we put that product out we were making the transition of allowing the Jamaican consumer to have a taste of the regular black tea that they had been exposed to in a tea bag, but then we steeped some

peppermint into it and that product is what we like to think is what created Jamaica's herbal tea business, because it was the bridge from traditional black tea into a product that had some indigenous materials in it. Over the years we have created cerasee, we have created lemongrass, which is fever grass, we have made straight Jamaican peppermint, which we call classic peppermint, we have also gotten into products like, in more recent times under a different brand which we call Sipacupa.[5]

we have started with moringa, which is really popular, guinea hen weed; we also have recently put out neem leaves, which is part of the Ayurvedic medicinal, an Indian/African Diaspora – that's where the root of that product is from but it's a very good product. Turmeric, which is yellow ginger, which is a very top notch spice. Turmeric is like a cousin to ginger and works well for people who have issues of inflammation.[6] ... And then we also are working on guava leave tea, which we hope to put out soon. But whatever indigenous raw materials and indigenous herbs that have medicinal value and benefits we are working towards bringing those products to the Jamaican consumer and to bring them more to export them to the various markets that we export to.

Product process and quality control

The production of teas does not involve an elaborate and complex process. While it requires a certain skill-set and great care to ensure product quality, the process is not a daunting one. It is clearly articulated in the response below from the owner of Perishables Jamaica, Norman Wright.

> what is involved is that they dry the mint, they reap it, the reaping and drying process can take anything from a week to two weeks depending on how they dry it.[7] Then we go down with our vehicle and we pick up the raw materials, the dried materials from them we bring into to our operations, and we sample it and send samples of it to the lab for evaluation to see if it has E-coli, salmonella, anything like that in it.
>
> To eliminate products of poor quality and then we grind them. At one stage we used to have that operation contracted out until we bought our own grinding machines – We then test the finished product to see that it is free from E-coli and salmonella because we have had issues where in the US we ship then our quality control wasn't as rigid as it is now. Products were shipped out to the US and then they would be dumped because they tested them and found E-coli and salmonella in them.

Having noticed that lack of proper quality control can hurt the business, Perishables Jamaica, is doing a lot to ensure that it has a consistently high quality product.

the intention is to bring the green herbs into that facility, wash them, dry them and come out with a consistent product that would be free from E-coli, salmonella and those other things. So, that's an on-going project and process that is taking place because we want to be able to do just that – to have a centralised drying facility. It is taking a bit longer than we anticipate but we are working through that process and knowing that the journey of a 1000 miles started the first trek.

Customers and markets

The company sells its products to both retail and wholesale customers. It also has an active export business in Europe, North America, and the Caribbean. Further, the company also does contract packaging for a number of other companies which are involved in the tea business as well.

> We export to the UK, we export to certain other markets – Barbados, we export to Cayman, we export to the USA under different brands, we also export to Canada. We contract pack products for other brands as well we contract pack for Ocho Rios brand, we contract pack with JCS – Jamaica Country Style brand, we contract package for Eden Gardens......, we contract package for Caribbean Dreams – some of their products, we also contract package for P. A. Benjamin under their Tropical Blends brand. So we offer various brands the ability to have indigenous Jamaican products packaged under their brand, which they in their own right distribute to their various markets globally as well.

The competitor landscape in which Perishables Jamaica operates makes for interesting reading as well. There seems to be a lot of collaboration and friendly rivalry among the firms even though they try to survive and prosper individually. This offers a very useful insight as to how the sector operates.

> We are unique in a lot of ways in that unlike our primary competitor Caribbean Dreams, which imports a lot of their raw materials, we use 95 per cent of our raw materials that are indigenous raw materials. So we are unique in that respect in that we used a lot of indigenous raw materials that provide work for Jamaicans. ... And so locally, we have that competition with imported products as I pointed out we do make some of the teas for Caribbean Dreams, which is Jamaican Teas Limited. We have a friendly adversarial business relationship as two small Jamaican companies, we

being the tiny one, and they the bigger one. We share our expertise and knowledge to some extent and, put it this way, we are friendly competitors if that is possible. And then we look more at the global markets in terms of exporting and earning foreign exchange. So globally we go to the world market with what we like to think are unique Jamaican products; like our peppermint. I don't know if you would find another peppermint like our Jamaican peppermint globally because its satureja viminea, which is an indigenous Jamaican peppermint. The peppermint that Grace, and Tetley and Caribbean Dreams and all of those other products that sell in Jamaica is a different peppermint from our peppermint. The flavour is different, so our product is unique in that respect. And our lemongrass and cerasee, while there are globally similar products based on where ours is grown and the species of it that we use is somewhat different in its flavour profile and taste, but as some have the similar chemistry to other products that have a similar name, you know...

We as a company [we] have about 15 customers – that's our customer base. We ship to two distributors in Canada, there is a third one that's not so active. Into the US we ship to two distributors of our own, and then we ship to two people we contract package for. All of them take the products to the various supermarkets. Some of them go into WALMART, some of them go into the Win-Dixie and Publix, into the chain stores under our brand and with their brand; and then like in Barbados, St Maarten, Cayman we just have one distributor we ship to and then they put the products into the market. In Jamaica we ship through one distributor CARI-MED Limited and they take it into the stores for us. That's it; we don't do the leg work we are not here for that. We are a manufacturer and we ship to distributors and they do the distribution for us.

While most small firms lament about sales and expansion in the export market due to liability of foreignness, this does not seem to affect Perishables Jamaica in a serious way. Instead, the most important thing for the company is making sure it has consistently high quality products. Once that is in place and it meets the requirements of the export market, then the sales is not a big problem.

> Once your product is of good quality, properly packaged in terms of the label requirements and the whole question of language like coming into Canada you have to be bilingual.... French and English you need to have that covered, the nutritional information, the bar code once you have those things in place and if it's free from any contaminants, the product sells itself in many instances.... So the product is really what does the selling for you, the quality of the product.

Corporate governance and employee relations

Perishables Jamaica, unlike many other small businesses, has a clear organisational structure and seems to operate in a similar fashion to large and well-established enterprises. There are five other shareholders in addition to the owner. This is not very common for firms of this size especially in the Caribbean context.

> There is a Board of Directors. I sit as the Chairman, we have a finance director, director, company secretary, we have a company lawyer if we have any issues, then we have a supervisor, our special projects manager, our quality controller, our machine operators, and the slew of our persons who take care of the sanitation of the compound and the sanitation of the staff clothing... We review our business plans every year and upgrade it and out to make sure that whatever it is that needs to be done we try and achieve them, what is not achievable we put it back on the drawing board or we dump it.
>
> The Board meets regularly.... we have an annual general meeting every year and two of the Directors are active in the company along with myself, the Company Secretary works fulltime with the company, and the Director of Finance is at the company every week involved in any critical issues and decisions that need to be made. So it's not a one man band running the business.

While Mr Wright is integral to the business, he credits the success of the business to his staff. Staff welfare, training, and development are taken seriously and play a key role in helping the firm to not only survive but also prosper in the highly competitive marketplace.

> We provide a fresh set of clothes each day to staff so we buy the clothes, we get them washed each week, so every morning production staff come in they take up a clean set of pants and shirt and go to the factory..... In terms of the work environment we make sure that the ventilation systems are good. In terms of the management we make sure that we pay all our taxes that are due, our statutories on a monthly basis, our GCT, we meet all those commitments...... the staff members go to whatever HACCP training because we aim to try and focus on quality. All of them are trained first aiders in terms of health and safety. The Bureau of Standards Jamaica is always putting on courses so they are always opportunities for them to go on training like that. We have a young university food chemist with us, going about 2 years now who comes to the table with his own knowledge and expertise. Up until recently, we had a couple of young chemical engineers from UTECH who we bring in from time to time but the green lights of the other companies sometimes they move on... The qualified accountant been associated with the company for well over 15, 20 years, who has worked with international

companies like KIWI brands Douwe Egberts, Sara Lee also with Ericsson, who were DIGICEL platform base... Then we have employees that are there with us for 13 years, guys that are there over 10 years... we have two consultant electricians who work with us for the last 30 years; anyone of the two can back up the other. Every six months we are audited by the US Army, the food inspectors, we sell products to the Guantanamo US Army base.

Future of the business

> We have ventured out over the years into other business opportunities, one time we tried to make a conch soup... but when we went commercial with it there were issues with the quality that almost cost us the business.... From then we focus ourselves on just doing the tea business and keeping our focus there.

Those words no doubt, tell a big story as to where Perishables Jamaica intends to go as a business. With the clear and insightful leadership from its principal owner, the future of the company seems to be in good hands. The oversight from their Board of Directors, the enthusiasm of the staff, and the strong social responsibility which the company shares with its stakeholder communities are indeed big positives that will continue to ensure its survival and prosperity.

> So each year that you create a new product you pumping back the money into the business but you growing and you taking on more farmers, as long as we can break even, and we can pay our farmers, we can maintain our goodwill... we are ok.

The company also has great plans for some future developments that will enhance its competitiveness even further as well as its survival and prosperity.

> We just got approval for a grant from Caribbean Export Development Agency (CEDA). Once we can access the funding we will be able to get a machine that would be doing an envelope wrap which is going to take our product quality up one notch instead of just having a regular tea bag wrapped over, wrap in an individual envelope which would allow us now to put our products into the hotels.
> We are in the process of installing solar panels to generate solar energy for our operations... In another three months, we should be running most of our operations on solar energy during the sunshine days. Downstream, we

will look to sell some of the excess energy to the JPS system, so we intend to be self-sufficient and truly an independent Jamaican company (which is not dependent on foreign exchange).

One of the important big plans is to take the company public soon. This will be a big step for a small company such as Perishables Jamaica.

> And at the same time we are working as a company to create the framework to go public, we are looking actively at that. We have been looking at it for some time and I expect that by our 35th year we should be going to the market to get some additional shareholders in the company.

Financial management issues

From all indications, it appears that the company is managing its financial affairs well. There has been growth in revenue, expenses seem to be under control, and there has been strong compliance with statutory obligations as well.

> Last year ending October sales were 70 million Jamaican dollars, the previous year it was $51 million. And then this year we expect to do about $80 to 85 million this year ending October. It could get better but that's a conservative, realistic, achievable number we working with there. Our exports grew by 50 per cent last year. So far this year they are up by another 45 per cent but then when you take out the devaluation of the Jamaican dollar you might find that you really only have grown by about 20 per cent but growth is growth in any context. ... We don't make a lot of money at the end of the day in terms of net profits but our mission is to grow.

The doing business environment

However, while internally the firm is strong and employees seem happy with being a part of the organisation, the firm's external environment has to be taken into consideration when looking at its future prospects. The entrepreneur does not always seem happy with the developments in the business environment and laments some of the weaknesses which he sees as things that can lead to the lack of business growth, survival, and prosperity.

> I put it to you that based on the situation that prevails today where you have to pay JA $60,000 to be a company, there are a lot of companies that will never be born, and lot of ideas will be still born and are being thrown out in

the bath water…. You have to have J $100,000 to start a business in Jamaica at least, and that's just to start it not to make the first or second or third payroll.

Despite this pessimistic outlook, the entrepreneur remains positive about dealing with the vagaries of the business environment in which Perishables Jamaica operate. This is what he had to say:

> Well, you know, as a company we have always sought to manage our business within the constraints and confines and the idiosyncrasies of the government.

Concluding thoughts

Perishables Jamaica Ltd has had a tremendous journey from its humble beginning with J$300 in 1980 to looking at JA$85 million in sales after 34 years. Similarly, this tremendous performance has won them multiple awards. The owner won the award for pioneer from the Jamaica Exporters Association (JEA) Pioneering Industry for exporting over 30 years and also won two awards from the JEA Best New Exporter Champion Manufacturing. The company is a net foreign exchange earner for Jamaica and contributes significantly to employment especially in rural communities among small farmers. The country still has to grapple with issue of quality control in order to remain competitive in the export market. Also, it has to deal with the challenges of climate change given the vulnerability of its inputs to weather conditions. Importantly however; what is at the forefront of the entrepreneur's mind is the survival of the company.

Notes

1   University of Technology, Jamaica.
2   Exploring Business Potential of Herbal Teas, (2013), Marketech Limited, Scientific Research Council.
3   Now University of Technology, Jamaica (UTECH).
4   Pep-o-mint is a blend of black tea and Jamaican Peppermint.
5   "That's a new brand we created a couple years ago and we pushing ourselves towards looking at the products that deal with more medicinal products."
6   "I had a call of from one young lady, who told me that she call one day and wanted to talk to the chemist because she had something to share with them:

and she told me that her dad who had been in a wheel chair for how many years, she gave him, and in pain, she gave him the turmeric tea and that the pain not only had he been relieved of the pain but he was able to get up out of his chair and walk."

7. The farmers do the drying because of a lack of space at the Perishables Jamaica facilities. "We don't have the space, the space, the space. We are working with the, we formed a cooperative with our farmers in South Manchester, the South Manchester Herbs and Spices Cooperatives Society. We formed that about five years ago, and we've been working through those five years along with help from JSIF to establish a centralised drying facility".

7
Caribbean Flavours and Fragrances

Abstract: *Caribbean Flavours and Fragrances produces and sells fragrance to wholesale and retail customers who use the products in various food items that they produce such a sodas, syrup, and so on. The company started from a humble beginning and has become highly profitable; it carved out a space in the very difficult manufacturing sector in Jamaica. This case looks at the early beginnings of the firm, its expansion over time, and provides insights into the future of the enterprise. In addition, it focuses on the firm's products, production processes, supply chain issues, and quality control processes as well. Similarly, the case also looks at the governance of the business and the business environment within which the firm operates and extrapolates how this environment will impact on its future.*

Williams, Densil A. *Competing against Multinationals in Emerging Markets: Case Studies of SMEs in the Manufacturing Sector.* Basingstoke: Palgrave Macmillan, 2015. DOI: 10.1057/9781137500328.0010.

the flavour business is unique. And I say this because a lot people think of fruits and they think of juice. Flavours are different from the fruit and the juice. Flavours are generally extracts from the skin of the fruits or the bark of the tree or the leaf of the tree, and flavourists then use the extracts as their base or core to create a flavour that can be used for particular applications whether in beverage, baked products, confectionery etc. It is in this application that the true taste and smell of the flavours can be enjoyed.

Anand James, Managing Director.

Introduction

Who would have thought that a teacher of history and an avid cricket fan from Guyana would have become a celebrated businessman in Jamaica? Well, Anand James, the former Chairman, and now Managing Director of Caribbean Flavours and Fragrances has made this a reality. The story of how Caribbean Flavours and Fragrances was born and has transitioned to become a success story in a sub-sector of the manufacturing sector that is not well understood, makes for fascinating reading. Some would say it is sheer chance that led to the success of this company, while others would read it as the result of the strategic leadership and foresight of a determined leader. After almost 13 years of operating as an indigenously owned company, the successes of Caribbean Flavours and Fragrances reflect a story that will inspire other persons who do not have a business background but are interested in starting and growing their own firm.

However, the journey to become an entrepreneur was never easy and did not always have the support of family members but the true test of that entrepreneurial spirit is whether or not the individual believes in the conviction that he or she can succeed and converts that belief into action. These realities were evident in Anand James' case.

> when I saw the possibility of becoming an entrepreneur that is taking a risk; and my interpretation of risk; having an idea, take the risk and follow it through... A lot of entrepreneurs just have the idea, it has to be a follow through and the risk, the risk has to be there. And when I told my wife that I'm going to invest all our savings into this business and she said, "No way, you are not taking the children's money and putting it into the unknown ...".

Risk taking is almost inherent in Anand's blood as he has a family history of risk-taking, which may explain the birth of Caribbean Flavours and Fragrances.

> I am a risk-taker by choice, by nature. My mother was a buy and sell person. From my mother to my siblings are into risk-taking ... there are 12 of us – six boys, six girls. Four of them end up in their own businesses; two created their own business and two end up in a situation like mine, where they were working with a company they saw the opportunity, the company give them that opportunity and they grabbed it and they took that risk and they were successful, just like mine.
>
> So entrepreneurship is something you have decided and the opportunity never came up to allow it to flow, and sometime you don't have an idea at all an opportunity comes up and somebody just say you either take it or you may not get a second chance. I am convinced in Shakespeare's Cassius in Julius Caesar that says, "there is a tide in the affairs of men, which, taken at the fall leads on to fortune." [Omitted, all the voyages of your life is spent in shallows and miseries.]

Industry conditions

Despite the vagaries of the international economy, the global flavours and fragrances market continues to show signs of buoyancy with an average annual growth rate of 4.2 per cent. The total global market in flavours and fragrance was valued at US$16.6 billion in 2012 and is expected to grow to US$20.3 billion by 2017. The market is fairly evenly distributed between flavours, which accounts for 52 per cent of the industry and fragrances, the remaining 48 per cent. The major regions include North America, Asia-Pacific, and Western Europe in that order; with the 2017 projections forecasting that the Asia-Pacific will supersede North America as the top market. Scholars and policy-makers argue that the dynamism in these emerging and developed economies such as Japan, China, India, South Korea, and South East Asia has given rise to higher incomes levels, which has translated to the purchase of consumables.

When segmented, the beverage sector was the largest global end-user of the market for flavours in 2012, accounting for 34 per cent of the market, followed by the dairy industry with 13 per cent. Other notable sectors included confectionary (9%), bakery (10%), and savoury/convenience foods (9%). As it relates to the fragrances sector, soaps and detergents (32.3%) accounted for the largest portion, with cosmetics and toiletries (26.9%), and fine fragrances (18.4%) also figuring prominently.

As it relates to companies involved in the industry, Givaudan, Firmenich, International Flavours and Fragrances, and Symrise occupy the top positions, accounting for over 60 per cent of the global market.[1]

The case

Business origin and expansion

The birth of Caribbean Flavours and Fragrances in 2001, signalled a new direction and a commitment to local production by its owner Anand James. Unlike most other small businesses, which start with neither the necessary customer base nor market presence, Caribbean Flavours and Fragrances was a bit different in that it already had a parent company with a strong presence in the market. However, the parent was ready to depart and leave the child to grow into its own. Having the entrepreneurial drive, one would say, to improve one's lifestyle, Anand James did not miss the opportunity to acquire the company from its parent and then designed an audacious plan to move the business from a mere subsidiary of a larger organisation to one that is independent and locally owned in the marketplace. The owner recalled that the parent company wanted to create a trading operation in Jamaica but he had to fight "tooth and nail" for the manufacturing to remain here in Jamaica. The narrative below describes how the company was born as a locally owned manufacturing plant.

> we came here, myself and my wife in 1983. I finished my first degree at the University of Guyana, and we came here hoping to finish my second degree, which I did; and my wife also finished the PhD because we came here both as teachers. We went up to James Hill in Clarendon, to Claude McKay High School. We did a good two-year stint there, then we came to Kingston we spent another couple of years at Kingston Technical High. During that time I finished my masters.
>
> Both the degrees were in history. As I said, we love this country, we didn't think of going back. In 1989 I wanted to look around outside of teaching for obvious reasons – salary conditions, and I got a short stint at Morgan's Harbour Hotel. I didn't like it, and I was leaving, and the Chairman of the Board happens to be the Chairman of this Board. The company here was called Bush Boake Allen and I told him that if I didn't like the work I will go

back and teach. And he said, "listen, we have an opening at the factory," and I say, "okay, let me try it out." This company is a flavour and fragrance company and I used to teach History and English.

and it was a hell of a transition here. But in the factory, I came in and help to teach the youngsters in the factory basic hygiene, little Maths and a little English. And, we were doing so well that I learnt from them, they learnt from me and I moved up in a flavour company. I went back to school of course, I went back to UWI, I did operations management, safety, health, sanitation, HAACP, ISO; and I got overseas training from this company also because it was a multinational company. I moved up and became Chairman and General Manager so it was far away from where I started. So when Bush Boake Allen was bought out by International Flavours and Fragrances, they offered us to become a trading company; in other words, they will make the product somewhere in the world and send it to us. And I said, "no, I wasn't comfortable".

I could have accepted being an agent and distributor for this big multinational ... I don't have to buy raw materials, maintain inventory etc. ... However ... I am not comfortable, I like the idea of manufacturing, employing people ... but they were giving me a nice offer and dangling this thing. "You only need two people you know Mr James, you need a guy with a forklift, and you need someone to cut your invoices, and you go and sell; and all you do is that you sell 1000 gallons of ginger beer, and we will bring it to you and then you distribute it to the different people." And I said, "I'm uncomfortable with that I would love to make the ginger beer right here just like how I used to do it, and I want three people to be making the ginger beer." And so that's a risk in a sense, that's entrepreneurship and a challenge ... In the end I negotiated to buy it instead and continue the operations with all my staff. That's how Caribbean Flavours was born in 2001 ... I mean, other than that, other than teaching, this has been my joy.

Products, production, and quality control

As the name suggests, Caribbean Flavours and Fragrances is in the business of producing flavours and fragrances to companies to make their various products such as beverages, baked products, confectionery, pharmaceuticals, home care, body care, and so on. It is indeed a unique line of business, which most persons would think that a chemist rather than a historian would be more interested in joining. Below is a full description of the products made by Caribbean Flavours and Fragrance, and also, the process involved in the making of these products.

So, if you squeeze orange juice and you leave it in the freezer, and you take it out the next day, and you want drink it as orange juice and you allow it to become liquid again you will find that it taste different and it looks different; and so here is where the flavour from the peel of the orange comes in to enhance that smell and taste. And this is what we do for pineapple, orange, banana, everything; so that's where flavour comes from – out of the extracts.

We have artificial flavours also, like kola champagne or cream soda which we just make into its natural state but 90 per cent of flavour comes from the botanicals and plants, and roots and what have you. For instance, we make ginger beer flavour using the extracts of the ginger roots what we call the oils and the oleoresins.

In the process of extracting the flavour materials from these oils we also have the by product called turpenoids or insoluble, and that's what we use for our fragrance side. So if I sum up, you would take the oil from the lemon peel and we will take that in a process, we will separate – the soluble from the insoluble; the soluble part will go in to make a lemon flavour – so you can make a lemonade or a Sprite type or Seven Up type and then the part that remains, which cannot mix in water, we'll take that and make a fragrance that somebody, who wants to make dishwashing liquid will end up making dishwashing liquid smelling like lemon or lime or orange.

Production and quality control

While to the uninitiated, the process of extracting flavours from fruits and leaves seems complex, for Anand James, it is just an uncomplicated process. Despite the simplicity of the process, Caribbean Flavours and Fragrances has taken a deep interest in ensuring that the quality of the output is of the highest standards. This is even more relevant in the context of the increasing international presence of the firm through its export business. The narrative below best describes how the process unfolds.

The process is simple. We are a batch manufacturer, and it's all liquid flavours and fragrances, all liquid. We simply have formulas that are unique to the industry, so there is a formulation to make cream soda flavour or pineapple flavour or an emulsion, which are those products that has the cloud in them. If you buy an orange drink, you will see it's a little cloudy, a pineapple or a ginger beer, whereas if you buy a cream soda or a kola champagne it is clear, so one is considered essence, the other is considered emulsion. An emulsion has water and oil mixed together with other things, and so we will pull out a formula and we will hand the production people and that's based on an order coming in, and they will put these raw materials in a particular way, at a particular time, and at the end of it you have a liquid flavour that can be used to make a syrup or ice cream or whatever. Just to let you know that we are working towards HAACP, which is food safety management system. The hardest part is to get it right the first time. It takes as long as a year to create a flavour formula before it enters the market as a new product. There has to be market tests, shelf-life tests, etc. before the customer will create labels and launch.

Customers and markets

Caribbean Flavours and Fragrances sells its products to both the local and the international market. While they have small clients, their major sales are to large manufacturers. Indeed, Caribbean Flavours and Fragrance services some of the largest firms in the Jamaican market that also produce products, not only to sell to the Jamaican market but have a significant and thriving export business as well. Given the significant demand for the flavours and fragrances in the local market, Caribbean Flavours and Fragrance exports directly only about 12 per cent of their output while selling the remainder locally. Their export destinations are not limited to the Caribbean, including the Dominican Republic, but extend to North America and Europe as well. It is very impressive for a company of this size to be making forays into so many international markets.

> we are about 85 per cent, 88 per cent local and the rest we export directly to customers in Barbados, Canada, Trinidad, Guyana, St Kitts. In fact, we are exporting this weekend to the Dominican Republic, and we did export to Belize and Puerto Rico some time ago, but now we basically make it the English speaking Caribbean, and we are just breaking through the Dom Rep. local markets; we support some of the biggest companies in Jamaica, you are talking about D&G² line of sparkling flavours, you have ginger beer, pineapple, kola champagne, and cream soda. You are talking about your WISYNCO, you are talking about the juice manufacturers, the syrup manufacturers, the milk manufacturers, cookies, baked products. We even are into pharmaceuticals – the syrups, and confectionary too, we do stuff for confectionaries. So, we are into some of the biggest manufacturers both here and in the region. What we also do indirectly – export. Maybe if we check the indirect export, we probably do 50 per cent local and 50 per cent overseas export and what I mean by that is that we make flavours for a customer, a big customer in Jamaica here they ask us to make it and ship it directly to the manufacturer abroad. We don't take the credit for that, so in that way, it's an indirect [export].

> Then there are customers who ask us to manufacture and to ship it out for them. And then, there are the customers in Jamaica who buy our flavours and they let us know very carefully their requirements... because the overseas market is very selective and so, a flavour or a finished product that can go into England maybe cannot go into Canada. One that can go into Canada may not go into the United States of America, so we know the regulations. So what these customers will do, they will say, "Caribbean Flavours can you make a flavour for us because we are going to make a finished product, whether it's a syrup or soda or a baked product and to ship it to Canada so make sure that when you make that flavour it is allowed into Canada." And, they will take that flavour and they will make the product and ship it abroad. So that's an indirect and if you add those then we are way above 50 per cent mark.

Further, the entry into the export markets is not always effortless for Caribbean Flavours and Fragrances. The assistance of larger companies to get their products into foreign markets is a major boost for the firm's foray into exporting. This networking with larger and more established firms seems to be an important contribution to the success of most small firms especially in manufacturing. The story below, as described by Anand James, gives a clear indication of the significance of networking with larger businesses to get his products into difficult export markets.

> generally Jamaica has a name, it is a brand abroad, and when people who have left here; they are used to the crackers and the Jamaican sodas, and so when they are abroad they would look for those things. And the manufacturers of those products such as your PEPSI, your Grace Kennedy, or your WISYNCO they will now look at how to serve those markets. And what they have done is; some of them attach themselves to overseas co-packers. You have people in Canada, in the United States, in the UK, who would manufacture for a Grace Kennedy here, or a PEPSI here, or a WISYNCO and they will ask us, or come to us for the flavour because these companies will give them and say, "listen this is the formula I want you to make a kola champagne you put X, Y, Z and Caribbean Flavours special whatever it is," and so these people would get in touch with us. Those other customers who find a co-packer in these areas again UK, US, and Canada they may say, "listen we are going to supply you with all your raw materials directly from Jamaica," so they will buy from us for their co-packers.

Similarly, the strong brand of the former owners of the company has not hurt Caribbean Flavours and Fragrances' foray into international markets. In fact, it appears that the former affiliation with the parent company had helped Caribbean Flavours and Fragrances to better enter into some physically distant markets.

this company was previously a multinational, so we had been part of those areas when we were a part of the multinational so we ride on that. Typically, what we do if a customer sends a request whether it's through the Internet or whatever, they see us on the website or they heard of us, and they said we will do our investigation and see what are the requirements for that market. We typically tell the customer or potential customer what the requirements are for their market. Once you get it right the first time, you get it right every time.

Although Caribbean Flavours and Fragrances has been doing well, there is still a bit of competition from other local firms in the market. Its main competitors also originate from the same background as Caribbean Flavours and Fragrances. Anand describes the competitors' landscape well.

> We have a couple local competitors. Most of these competitors were also part of multinationals... about 20, 25 years ago the trend started whereby multinationals were pulling out from the Caribbean for different reasons – small market, lots of regulations, etc. And I think the realisation is that they can pull out their responsibility from the region yet maintain market presence.
>
> the multinational that owned the flavour house in Trinidad sold out; the three that were in Jamaica sold out, so in Jamaica you have Virginia Dare which was part of Virginia Dare United States; and you have Flavourland which was part of Givaudan Worldwide and then you have Caribbean Flavours, which was part of Bush Boake Allen Worldwide. In Trinidad you have Tastemakers and Stewart Brothers who were part of Givaudan.

The survival of Caribbean Flavours and Fragrance rests on the vibrancy of manufacturing in the local market because a large portion of its outputs are sold locally. The details below show how interconnected the firm is in the fabric of the manufacturing sector. This may have both negative and positive effects on the longevity of the firm. Critical to note, is how the openness of the local economy, which drives up a lot of imports, has also made business difficult for Caribbean Flavours and Fragrances.

> Generally, flavours and fragrances are really raw materials for your home care, body care and your food, right. The more we import a finished product the less will be the demand for our flavours and fragrances. So, if we are bringing dishwashing liquid from Dom Rep or Costa Rica, immediately you have pulled us out of the possibility of being a supplier, and if you deal with food, all the snack food that's coming from all over the world that robs us of the potential of the market. So the more we make these things here, whether it is hand lotion, hand cream, body lotion, dishwashing liquid, soaps, toothpaste,

candies, sodas, snacks, if you make them here we have an advantage because we can knock on the door of the manufacturer here and say we can provide your flavours and fragrances need.

Once you have a manufacturer here then we are in a good speed and so unfortunately, we don't have many manufacturers here. Every now and again a new product would come out... if it's not made locally we don't have a chance. So that's if you ask the market is kind of constricted by how many players are manufacturing locally. That's why we want more manufacturers here and once we have manufacturers we have a chance to sell our products.

Corporate governance and employee relations

Caribbean Flavours and Fragrance is not like many small businesses that are solely dependent on the owner to get things done. While there is an indication that the owner has a strong say in decision making, he still has to multitask to keep the head count at a level that is manageable for the size of the organisation; hence, this may explain the company being well structured and operating along the lines of well-established, larger or multinational enterprises. It has a Board of Directors that is independent from the ownership of the company, and a clear organisational structure is in place with roles and responsibilities clearly defined. This arrangement might have been the result of the company's decision to become a listed firm on the Junior Exchange of the Jamaican Stock Market. The owner has been very bullish and enthusiastic about this move.

> we now have a Board that have independence – some of the biggest names, Mr Howard Mitchell who used to head NHT, he is our Chairman; you have Mr Clive Nicholas who was head of Taxation he is here, which gives us insights into taxation issues; Billy Heavens who heads Jamaica Cricket Board and also CHASE, they are wonderful people. They have accepted to serve on our Board out of friendship and out of a commitment to country and entrepreneurship. They are very cognisant of the critical role this company plays in the manufacturing segment of our economy and so we went public, and we have done very well in terms of organising ourselves structurally and making better business plan and long-term plan. Whereas, when you were a privately owned company you make decisions based a lot more on gut feeling, on your relationship, now you make decisions on a more sustained basis.

There is also a clear need that staffing will be important in order to make the new governance structure work effectively. The company, while managing its complement of staff well, realised that to effectively build a productive and competitive organisation to compete internationally, it would require a greater focus on building talent and organising the workforce well. This is clearly not lost on the owner and managing director.

> we are now bringing in a general manager, so for the first time you have a company with a managing director and a general manager. I am giving my full support, but the leg work now is going to be done by a general manager and a more youthful person who can run with it. After 25 years in an industry like this you tend to get tired at the leadership level you know, and so we went public, we now organise ourselves, and so we are going to expand. That's how it is.

While still trying to build the staff complement, however, there is great care taken to ensure that the strategy is not merely based on numbers but more aligned to the organisation's strategic vision and financial management. It appears that employment in the small business requires the ability to multitask as this helps these firms to better deal with cash flow issues especially when operating in a highly competitive marketplace where demand can be low.

> We have about 12 personnel, very small, very compact. Again, I always say you have to watch your cash flow; cash flow can bring you down. In a market condition like what it is over the last two to three years, where people are extending their credit to 60 days and you have to pay your GCT after 30 days, it can be tough. So you have to watch your cash flow. How do you do that? Make sure your head count is tight, small. That comes back to your staff. If they can produce for you; you don't need to employ 13 when you can employ 12. As I mentioned before, in a small organisation, there has to be people who can multitask, who can do different things that helps their earning power also.
>
> As a small business turns you cannot afford all departments. You have a lot of multitasking, for instance, you have a single person who does regulation, who does all government arrangements and relationship, who will be the same person who do your marketing, and the same person who do your human resources and you have a person who is customer service, sales manager, purchasing, chief cook and bottle washer.
>
> I have quality manager that also act as a plant manager and they take on different, different approaches. In small business, you want to pick people who can multitask. Government regulations can be a nightmare. There are various regulatory bodies whether it's Trade Board, Bureau of Standards, the Pharmaceutical Services Division, or the Ministry of Health or Labour etc.

DOI: 10.1057/9781137500328.0010

Critical for Caribbean Flavours and Fragrances is the need to have staff well trained in all areas. The unique and cost effective ways in which the company carries out its training makes for very interesting reading.

Caribbean Flavours is very big on training. Actually yeah, jokingly, my new General Manager looked at the file and said, "Mr James you have been training for export, because we take people out of university and college and high school, we send them to Bureau of Standards, we send them to SRC, we send them to UTECH, we send one to UWI and then when they picked up the experience, we send one or two of them abroad and then our customer woo them away." When we lose them, we lose all of them to our customers so we don't feel too badly. And when they go there, they look at the operation and they say, "listen, I think Caribbean Flavours can help you out here." So we end up benefiting in a sense but training is a big thing for this company – two reasons, one, the nature of the operation require basic food safety awareness, secondly creativity and consistency demands training and exposures.

The other thing is that I am a teacher by heart, I teach every day. So when I walk around I see something, I said you know what you can do this a little differently. And then I realise that our people may be doing the thing the wrong way, I said let us solve this with some training programme formally. You [are] teaching them life skills also because some of our workers here have not gone to high school. Basic skills of weighing and pouring are not enough so you have to give them life skills also.

Everybody has to get one external training for the year; that's a rule. And generally, it's Bureau of Standards and it has to do with either HACCP food safety or good manufacturing practices. And then we have our internal training once per month and this is scheduled, and so it's known there; I think one hour last Friday of each month, it's scheduled and the Quality Manager will take them through safety in particular; sanitation in particular; hygiene practices; and so on or use of equipment; use of safety gears; etc.

Future of the business

Close to Mr James' heart is the survival of the business long after he has exited the stage. This seems to be one of the major motivations for him to have listed the firm on the Junior Market of the Jamaica Stock Exchange. Further, the uniqueness of the business and the need to maintain the true Jamaican flavours of a number of the products, which his line of business has contributed to over the years, is critical for him. There is no doubt therefore that the planning, strategic foresight, and dedication to staff training are all part of this greater plan to ensure that the firm continues to survive in a very competitive and dynamic marketplace.

One of the biggest things that this company did was to go public about six months ago, and the reason, we wanted to ensure that the company continues. My biggest fear in the 25 years I have been in this industry, about 19 of that as head of the organisation, and about 12 of that as head of my own organisation; one of my biggest fear has been if something goes wrong, our reputation will die. Because you are talking about flavours and fragrances that are embedded into some of the major products in this country. And, when I think of a Jamaican ginger beer or a Red Label wine, I said listen, if you can't give these people their raw material exactly, and they can't get it tomorrow you have embarrassed yourself, you have let down our customers. The staff is very aware of the critical role we have in the nature of our business. And the last thing you want to know is that D&G has to be bringing products from abroad to make a Jamaican icon product like a ginger beer, pineapple, cream soda, and so on.

Interestingly, although the company went public, the funds raised from going public still have not been used to carry out business investments in the short term. Clearly, the prudent financial management of the firm over the many years has put it in a good financial position so that the excess cash it gained from the public offering can be used for future developments of the business, not to settle immediate financial needs. This can have both its advantages and disadvantages.

> We went public in October and the money raised is going to be used for expansion, not for paying down debts as we did not have any serious debt. We micro manage our financial activities to make sure our cash flow is excellent.

The doing business environment

However, despite the very noble intention of ensuring that the firm continues not only to survive, but prosper in the competitive manufacturing sector, it can be extraordinarily difficult to operate in the business environment as a small firm. Having had the experience of operating at high levels in a multinational corporation, Mr James was exposed to the marked differences in how small businesses are treated compared to large and multinational firms as it relates to accessing financing, dealing with government regulations, seeking markets among other things. The narrative below eloquently captures the perspectives of the owner on how difficult it is to do business as a small firm.

> Small business is not easy, and coming from where you are looking at small and medium-sized business and I will give you a classic example. I could have as much a loan as I wanted from the banks when I was a multinational company and as soon as I became Caribbean Flavours, locally owned, the guy said your equipment is kind of unique, we can't give you much collateral therefore if you want to borrow you have to find something more tangible: and, those collaterals were acceptable just the day before because it belonged to a bigger company and a multinational.

General discourses in entrepreneurship suggest that part of the difficulty in doing business as a small firm, lies in the customers' lack of faith in the ability of the firm to deliver quality material on a timely basis; you have to prove yourself in order to survive. This was evident in the case of Caribbean Flavours and Fragrances as well.

> listen you are a small company, we were buying from you when you were a multinational, they had insurance in case anything go wrong, they have technical support I don't know what you are; you are a general manager that's what you are. You are starting this company, you want us to put our lives into your hands and suppose anything go wrong, what guarantees do we have? And I say, listen, I have been making it here when we were a multinational and we certainly can do it as a local firm.

While external factors have also affected doing business in general, the owner gave a good anecdote of how lack of proper planning can lead to a downfall of many small businesses as well. This was quite sobering and has to be taken as part of the cultural dimensions of business failure, which most persons tend to ignore. The narrative below captures this well.

> we have a guy here, who used to come here and buy fragrance to make small air refreshener for cars: he did so well in the first six months. I nurtured him, I said come make sure you start buying in bulk, benefit from the budget. The next thing the man came with a F150 one day in the car compound, I didn't know. I was walking to some region and I went out and I say, "What is this?" He said, "Anand, I get to sell you know, I mean, I get things in the back and so ... " I say, "a pickup would have been alright. Careful you won't carry down this business" – so said, so done. The next thing, I see this man in a taxi come to buy goods and I call him out and I brought him in here, and I said, "What happen?" He said, he over extend himself and thing and this is happening now he wants credit. I said you could have gotten credit when you were ahead of the game now I am reluctant to give you credit because of this. Eventually I nurture him back and he has now bought a small vehicle and he is doing exactly what he was doing without attitude, fanfare.

And surprisingly when I talk with my staff about that behaviour they say a lot of our customers are like that, they come here they start a bag juice business make a few sales and they think its profit. And I have a guy who after three years still coming to buy one, one gallon. I say "when are you going to buy a drum?" and he say, "Mr James no money nah run," and I say, "You not thinking big".

Concluding thoughts

Mr James has big plans for Caribbean Flavours and Fragrances as he is intent on ensuring that the business does not only survive but prosper for a long time to come. Market diversification is in the mix as one of the strategies.

Going forward we are looking at expanding into the Caribbean.

Starting in business however is not the most effortless thing for a small firm such as Caribbean Flavours and Fragrances. The business environment is tough, the market is highly competitive, and with an intolerable culture to failure, it makes it even more difficult for small firms to be motivated to stay in business. However, despite these challenges, clearly, Caribbean Flavours and Fragrances sees a ray of hope and is still willing to make investments to ensure the survival and prosperity of the firm. His parting words provide strong motivation to persons who want to get involved in business and build firms that survive and prosper.

Do not be afraid to TRY. Our biggest fear seems to be fear of failure. This can be a positive motivation as the mentality drives one to be careful, to work extremely hard and at all hours, and to be frugal in life style at least till the business can afford you a better lifestyle. The down side is that you may pass up an opportunity for growth or expansion because of that very fear. I have had my fair share of success.

Notes

1   "An Overview of the International Flavours and Fragrances Market", 8th Edition – IAL Consultants (2013).
2   D&G stands for Desnoes and Geddes.

8
Conclusions and Lessons Learnt

Abstract: *The Conclusions and Lessons Learnt chapter provides a synthesis of the material covered in the text and highlights the common lessons that lead to the survival and prosperity of all the small firms that participated in the study. The chapter puts forward the following as common lessons that have resulted in small firms that compete against multinationals in the manufacturing sector from emerging economies, surviving and prospering as: networking with larger enterprises, brand building, intimate knowledge of products and line of business, diversification of markets and meeting customer needs, and forward thinking and strategic leadership. These important attributes can be replicated across all firms and should be encouraged if smaller firms are to survive in an inhospitable sector.*

Williams, Densil A. *Competing against Multinationals in Emerging Markets: Case Studies of SMEs in the Manufacturing Sector.* Basingstoke: Palgrave Macmillan, 2015. DOI: 10.1057/9781137500328.0011.

Competing against Multinationals in Emerging Markets: Case Studies of SMEs in the Manufacturing Sector offers important insights into how some small firms have been able to move from their humble beginnings into established enterprises; prospering in the business, and just not merely getting by despite the inhospitable competition from large and multinational firms. The work presented in this volume drew on case studies of firms in the manufacturing sector to tell the story. The purpose of the volume is to provide common lessons that can be adopted/adapted by other small firms that are operating in highly competitive sectors with large and multinational firms and are looking for ways to survive and prosper. The manufacturing sector was chosen because it is seen as one of the sectors where small firms, especially those in the Caribbean region, will find it hard to remain open given the less than benign operating environment in most countries (high cost of doing business: high interest rates, unstable exchange rate, high cost of energy, difficulty in sourcing raw materials, difficulty in sourcing qualified workers, overcoming high levels of bureaucracy in government, and so on), and the openness of those economies to international trade. The Caribbean region is home to some of the smallest economies in the world and in most cases, some of the most open economies in the world; with trade accounting for more than 100 per cent of their GDP. The firms in this region therefore, are always open to competition whether from imports coming into their markets or from their engagement in international business through exports into foreign markets. If these firms are not internationally competitive, it will be quite difficult for them to remain open over any long period of time. This is even more so for the smaller owner-managed firms, which have a limited portfolio of products and also markets, and in most cases are unable to gain production and distribution efficiencies through economies of scale and scope. So despite all of these handicaps, why have some of these enterprises remained open, survived, and prospered while similar others, facing the same market conditions, have either died out or are just getting by? The lessons presented in the next section which are drawn from some of the most successful small firms in the manufacturing sector across the region, will provide some insights into how this might have happened. These insights will make a modest advancement in our knowledge on the difficult path small manufacturing firms have to take in order to survive and prosper in a difficult environment and especially against larger and multinational firms.

Common lessons

From a detailed analysis of the narratives presented, a number of lessons were observed on how the small manufacturing firms across the Caribbean were able to move from initiation to established enterprises, surviving and prospering, and not simply getting by. This section speaks to some of these lessons, which can help other enterprises formulate strategies to enhance the chances of their firm's survival in a hostile market structure and industry sector such as manufacturing. Below is a list of factors that are found to be critical for the survival and prosperity of the small firm in the highly competitive manufacturing sector, where larger firms have a clear competitive advantage and are logically expected to survive and prosper.

Networks

The common saying: "no man is an island, no man stands alone" is even more relevant to small businesses in a competitive market structure and industry sector such as manufacturing. A close reading of the narratives on how the firms reported in this study have been able to become successful enterprises from their humble start up shows that both business and social networks were critical in assisting them to move forward. Entrepreneurs reported on their association with business networks such as chambers of commerce, which helped them to gain proprietary information on the market as well as opening doors to very influential persons, who could help them to overcome the bureaucracy of government. In the Hot Mama's case, for example, the principal spoke about her husband's role in the Chamber of Commerce in Belize as an important source of brainstorming and getting business ideas, which otherwise would take a tremendous amount of resources to accomplish. These free ideas milled from one's network is not something small businesses should underestimate, as otherwise, they would have to pay a heavy price tag to consultants to get the same knowledge.

> here in Belize, ahm, he [husband] became a charter member through the Rotary Club in San Ignacio, and once a year they have a spaghetti dinner that they do in the market; sort of like promoting Rotary with people. And they were doing the prep work the day before, you know, having a few beers, they were cutting up vegetables for cooking, stuff like that. And the gentleman said

to my husband, "what's with this name Food Limited? It is so boring." And so my husband looked and said, "well, I know but what do you think?" He said, – "a real Belizean name, it sounds like a Hot Mama's," and that's how we got the name.

So he came home that night, a little tipsy and all that and he said, "Wilana, we are changing the name to Hot Mama's," and I was like, "Oh, okay".

Similarly, the principal of Island Moldings spoke of his connections with government officials and the company's sponsorship of local events, which gave him priority access to persons who could help to open doors for his business.

> the Government I don't have a problem specifically, they support me a lot ... We at Island Moldings, one of the thing we have done, once we make profits we support the Government for example, like the Police we donated stuff, anything that we can do for schools and hospitals, we always supportive of them.

Further, networking with other larger businesses is also useful for small firms to note. Some of these successful entrepreneurs have noted that without larger firms giving them a big contract early in their life cycle, it would not have been possible for them to survive and prosper. The Kittitian Hill project for Island Moldings, the Grace Kennedy project for Spur Tree Spices, the contract packaging done by Perishables for larger companies such as Lasco and P.A. Benjamin, Pepsi purchasing from Caribbean Flavours and Fragrances are important examples of how larger firms have supported smaller firms thereby ensuring their survival in a competitive marketplace. The example below speaks to one of the cases where networking with larger businesses was a strong factor in helping a smaller firm to survive and prosper while they competed in the marketplace.

> a very good friend of mine when I was in the restaurant business, she was in charge of Grace Foods Services Division. So because of the restaurant she used to buy. Plus from Island Grill days she knew me. I said to her, "why can't you sell my seasonings to the hotels?" and she said, "let me put in a bid, because they will open up that new market".... the rest is history ... we got the Grace Kennedy deal and that started that business there.

Another important example of how businesses work together to enhance the survival and prosperity of each other, especially the smaller firms is narrated below.

> And so locally, we have that competition with imported products; as I pointed out we do make some of the teas for Caribbean Dreams, which is

Jamaica Teas. We have a friendly adversarial business relationship as two small Jamaican companies, we being the tiny one, and they the bigger one. We share our expertise and knowledge to some extent, and ahm, put it this way, we are friendly competitors if that is possible. And then we look more at the global markets in terms of exporting and earning foreign exchange. So globally we go to the world market with what we like to think are unique Jamaican products; like our Jamaican peppermint.

Besides business and formal networks, an important network that has also helped most of the small businesses in this study to survive and prosper in a difficult sector is that of family. Family networks and connections have proven in most cases to be a big source of advice as in the case of Island Moldings and Spur Tree Spices, Yono Industries, and also a source of financial capital as in the case of Hot Mama's Belize. These sorts of more informal networks are useful to help firms either minimise the cost of capital for business growth and expansion or for a source of advice on business processes and management issues. Otherwise, it would be an exorbitant cost for the business as they would have to go to the open market for capital or hire consultants to provide advice on management practices. Most small firms would not have the requisite resources to undertake these activities, and therefore, if left only to the formal market mechanisms, they would not have survived long beyond their founding.

Brand building

It is not uncommon for most small firms to think that it is ok to survive in the marketplace by servicing their small pool of customers without paying attention to how the rest of the market sees them. Building their brand in order to attract new customers is not always at the forefront of a small firm's agenda because they see it as an expensive endeavour, which may not deliver immediate benefits to the business. However, this is far from the truth. Making your brand known in the marketplace is both necessary and critical for business survival of all sizes. Giving the brand greater prominence in the marketplace will help to attract customers as Island Moldings demonstrated with the Kittitian Hill project, attract investors, and also, get the firm into networks that it would not have gotten into. One of the successful entrepreneurs from the study sample summed up brand building nicely when he articulated that:

But you also wear the brand. We learnt this, everywhere you go, brand yourself. Whether it starts conversation on airplanes, in supermarkets, in banks, everywhere you go, it starts the conversation. You don't have big marketing funds when you start off, so you have to believe in the brand, you have to wear the brand, and show how enthusiastic you are. It will rub off. If you don't do that and you sit there in your office here and expect it to sell...you expect the caterer, you expect the distributor to take your brand new products with no image in the market it has gone into and expect to sell off the shelf? Unless it is half the price, the other ones up there won't sell. So you've got it the hard way.

The important lesson is that small firms have to invest time, resources, and effort in making their brands known in the domestic market, and also try to use that to leverage international markets as well. Brand building should not be seen as an expense for the business but as a strategic investment that will eventually lead to long-term benefits, survival, and prosperity of the firm. Too many small businesses look at the immediate expenses associated with brand building and see it as an added expense that will impact the business bottom line in the short term but they have not considered the long-term positive impact. Formal business networks such as social clubs, business meetings, using the media in a judicious way (provide content for media while at the same time getting recognition), social engagements in communities, and non-governmental organisations, and so on, are low cost ways of building the brand of which small firms should take advantage.

Intimate knowledge of products and line of business

Getting into business and ensuring its longevity is not merely about having an idea and then finding a person to execute these ideas on one's behalf. For the small firm that is resource constrained, and in most cases cannot afford to hire high level management expertise to carry out the task of executing business ideas, the owner has to ensure that [s]he understands the business in a very intimate way in order to efficiently execute tasks as they become due. In almost all the cases reported in this volume, one of the common elements that the entrepreneurs spoke about is their understanding of the business they are in. Most of the owners of these small firms started their firm based on their own previous work knowledge and expertise or they developed an idea based on their own academic pursuits. This was most evident in the case of Yono Industries for example. The owner designed his own energy-saving system, which delivered a competitive advantage to the firm. In the case of Island Molding, the owner was able to make the moldings, windows, and doors himself, and as such can provide better oversight to staff. In the case of Spur Tree Spices, the owner was in the food industry for a significant period before launching his own food business. With their intimate knowledge of their products, the small firm owners do not have to depend heavily on expensive management hierarchies, which in most cases, would not be available to these firms but are in larger organisations.

One entrepreneur sums up the importance of having an intimate knowledge of the products and the business in order to ensure survival and prosperity:

> The good thing with me, when I first started out, is that I as the owner/manager, I can do everything in this shop as these guys; and once I bring some guys on board basically I start to feel them out because it's not everybody that has that skill set. I mean they are guys who are bright, very educated but they just don't have the hands-on, they can't do it. There are other guys who just don't have an education but they are really, really good at hand, and so you need to find out these guys and once they really good on the hand – for me, I think with anything that you are going successful with – one has to understand how any product works and I think that's one of the key things, even in school, I remember going school and not understanding a lot of stuff is after I finish school I actually started to understand what I was being taught. Also, anything that you going to do, you got to love it because I love what I do...I enjoy what I do, and that's real key to any business.

Market diversification and meeting customer needs

One of the hallmarks of these successful small firms that are surviving and prospering in the various sub-sectors of the manufacturing sector, is that they not only depend on a single market in which to sell their products but they try to sell into various marketplaces. This finding is in line with the limited portfolio argument espoused by Hall in trying to explain failure among small firms. The small firms in this sample would have tried to overcome the limited portfolio stigma by diversifying their markets thus leading them to depend less on one single location from which to derive their revenues. While export sales are not always high in most of the successful companies, they try to have a foothold in international markets gained mostly through personal contacts of the owners. These sales have helped the companies to overcome many of the cash flow challenges, which they continuously face especially from domestic customers who take a long time to pay. It is important to note that all of the companies in the sample have had sales to international markets, whether through direct exporting or through their unsolicited orders motivated mainly from their contacts abroad. Some of the companies have also set up an online store to facilitate the sale of their products to international customers as well. This is something that most small companies lack; especially those in the Caribbean region. Their visibility on the Internet is woefully inadequate, and so, they lose out on major market shares especially among a growing segment of the market that has become keen on online shopping. The companies that have survived and prospered have shown that E-commerce is indeed an important part of the marketing mix to ensure continued growth and expansion.

Similar to market diversification, all entrepreneurs noted the importance of having satisfied customers to ensure longevity in business. This is even more important in the context of small firms, which do not have a myriad of products to offer in order to give them a large customer base. In most cases, these small firms have one or two large customers who are the lifebelt of their organisations. Having these customers dissatisfied would lead to a loss of business and eventually the closure of these firms given their very limited portfolio of customers. It is not surprising therefore that the entrepreneurs in these firms spoke highly of the need to ensure that they keep their customers satisfied. For them, a satisfied customer is not merely about customer service and having a smile on one's face when the customer walks inside the business place; it is about creating a product that gives the customer strong utilitarian value. This will require speaking with the customers to better understand their needs and then work with them to design a solution which will give them the greatest value.

The importance of keeping satisfied customers and an increase in the portfolio of markets to ensure the survival and continued prosperity of the small firm is captured well by two of the successful entrepreneurs in this study:

> With anything that you have done, or you are doing, there are some key factors that I think is really, really important that will make your customers happy and want to pay you ... Ahm, it's delivering on time, delivering a quality

product, 99 per cent of the time they are not going to give you the problem. Most of the time problems come in when you don't give a good product, you don't deliver on time then that customer get very unhappy, and it creates problem. So I think if you have a good product make your customer happy, and your client happy and I think that keep you getting pay on time.

The other thing we really ... we've done over the period is not simply targeting only the Diaspora in the major con-urban areas of the US or Toronto or the UK; because that is a shrinking market to some extent. It's a very price driven market, it's overly competitive. There are brands from everybody in there. But the vast continent of North America is looking for taste, new flavour, new seasonings... So that has opened up a huge market if you go after it the right way. So you need still to service the Diaspora market because you need to fill your containers up to make sure you have sufficient volumes to make it worthwhile. But the biggest market untapped out there is the non-Caribbean, non-Diaspora market in many ways, like in food service, in restaurants, that's in retail products in maybe the higher end, natural whole food stores that have to market, there is a massive market untapped, we can never begin to fulfil the whole of that market as a country but we have a good chance given the reputation of jerk and Jamaica has a good chance of doing it. And it is those opportunities which represent by far the largest opportunities.

Forward thinking and strategic management

At the centre of the small firm is the owner/manager. [S]he makes almost all the decisions related to the operations of the firm. Even though there may be consultations and advice sought from others, the final decision on the execution of any project or change in business processes generally is left with the owner in the small firm, unlike large and multinational firms where decisions are generally made by teams and approved by a Board of Directors. This may explain why when failure occurs at the small firm level, the owners generally see it as "persons" as they find it hard to separate the business from the person. Since the owner is at the heart of the decision making process in the small firm, it is therefore crucial that [s]he is forward thinking and pays close attention to developments in the global arena that will either pose challenges to the operation of the business or present opportunities that can enhance the growth of the business. Operating in a highly competitive marketplace and industry sector requires not only operational foresight but strong strategic thinking about the future directions of the enterprise in order to enhance the probability of the chance of survival.

All the firms in this study have displayed forward looking and strong strategic foresight, similar to their main principals which in all cases is the owner. The owners of these enterprises from origin, have never been satisfied with their enterprises remaining in the birth stage of their life cycle. They have always had at the forefront of their minds, how to move the enterprises from birth stage to maturity. This is evidenced in the number of deals that owners of these firms are trying to put into place to increase the likelihood of the growth of their firms. It is this type of forward thinking that has led to the success of these firms in the main, as owners have always tried to find ways of meeting their customer needs, ensuring that employees are satisfied, and efficiently manage their cash flow to ensure that they continue to meet their obligations in running the enterprise. The focus on building a professionally run organisation with the required levels of skills and competence is also crucial for the survival and prosperity of the small firm. Owners of the successful firms identified in the sample for this study have articulated the need for new strategic thinking in order to ensure the continued survival of their firms.

To sum up the critical importance of the need for forward thinking and strategic management to increase the chance of survival of the small firm, one owner puts it this way:

> put it this way... where we are now, we can't continue along the lines we are going. That is why we are trying to bring in an equity partner. We have to find the management skills, we have to get a right production manager, somebody with a degree ... we want to eventually have somebody who can grow into a general manager.... that is why we are putting a plan together. Because ... to do that, you have to increase the business to find more business to pay the person ... But we've reached the stage where we realise that potentially it's so that we are not going to be able to achieve that or grow at the rate we want to achieve and want to grow at, or should be able to grow at, unless we strengthen the management in all aspects, and financial management, accounting management and staff management, production management, and planning in terms of production and planning in terms of export marketing and growing of our production and exports.

Concluding remarks

This volume has provided copious evidence from the owners of highly successful small firms in the manufacturing sectors across the Caribbean region, a sector, which is characterised by high levels of competition from imports and also local production, high energy cost in most locations, high interest rates on the cost of capital, and an unstable exchange rate in some locations; all factors that mitigate against a successful operation of the enterprises especially the smaller ones. The evidence provides insights into how the owners of these successful enterprises were able to overcome the challenges in a sector and market structure that is not the most hospitable to their operations. From a careful analysis of the data, it was revealed that business and social networks, branding of the enterprise, intimate knowledge of the products being offered and the line of business one is in, having a portfolio of markets and a strong focus on satisfying customer needs, and strategic and forward thinking leadership are all factors that have helped these firms to survive and prosper in the sector and not merely exist. However, going forward, while these firms have operated successfully thus far, there are still some areas that they will have to strengthen in order to ensure their continued survival. There will be a great need for improvements in corporate governance of these enterprises. While they have reached this far in their operations without strong oversight from a Board and also having in place, a structured organisational framework for oversight of the business, as the markets become more competitive and these firms seek to expand and raise capital externally, the need for better governance will become even more urgent.

Similarly, in most of the production outlets for these firms, there will be a greater need for proper plant and facility layout in order to maximise the efficiency gains from their operations. In some cases, the production outlets are not properly organised and there is no logic to the flow of production. This led to inefficiency in inputting raw materials and also in the packaging of final products. Efficiency gains can be made with a more organised production outlet. Funding from multi-lateral financial institutions such as the Inter-American Development Bank (IADB) and also regional bodies such as Compete Caribbean can be sought by the entrepreneurs to acquire grants, which can be used to hire consultants to guide this process.

While the research in this volume aimed to be comprehensive as best as possible, clearly, given the limited resources to carry out the project, all countries in the region could not be covered. Future researchers should seek to address other countries in the region on this very important topic. Further, other sectors such as services, which have a high concentration of small firms, could also become a subject for study and the results compared in order to better generalised factors that contribute to survival and prosperity in the small firm sector.

It is hoped that the work presented in this volume has advanced our knowledge of the subject in a modest way. Understanding the factors that contribute to the survival and prosperity of the small firm should be of interest not only to small firm owners but government policymakers and persons interested in providing training to the small firm sector as the sector in most economies, especially the small, open economies of the Caribbean region is the lifebelt of employment creation and GDP growth.

Bibliography

Ahmad, N. & Seet, P., 2009. "Dissecting behaviors associated with business failure: a qualitative study of SME owners in Malaysia and Australia." *Asian Social Sciences*, 5(9), pp. 98–104.

Aldrich, H. & Reese, P., 1993. "Does networking pay off? A panel study of entrepreneurs in research triangle." In N. Churchill, ed. *Frontiers of Entrepreneurship Research*. Center for Entrepreneurial Studies, Massachusetts: Wellesley, pp. 325–339.

Ayyagari, M., Beck, T. & Demirguc-Kunt, A., 2007. "Small and medium enterprises across the globe." *Small Business Economics*, 29, pp. 415–434.

Baines, T.S., Lightfoot H.W., Benedettini, O. & Kay, J.M., 2009. "The servitization of manufacturing - A review of literature and reflection on future challenges." *Journal of Manufacturing Technology Management*, 20(5), pp. 547–561.

Baldwin, R. & Evenett, S., 2012. "Value creation and trade in 21st century manufacturing. What policies for UK manufacturing?" In D. Greenaway, ed. *The UK in a Global World – How can the UK Focus on Steps in the Global Value Chains that Really Add Value*. London: Centre for Economic Policy Research.

Barney, J., 1991. "Firm resources and sustained competitive advantage." *Journal of Management*, 17(1), pp. 99–120.

Bassi, A., Tam, Z. & Abi, A., 2012. "Estimating the impact of investing in a resource efficient, resilient global energy-intensive manufacturing industry." *Technological Forecasting and Social Change*, 79, pp. 69–84.

Campbell, N., Heriot, K.C., Jauregui, A. & Mitchell, D.T., 2012. "Which state policies lead to US firm exits? Analysis with the economic freedom index." *Journal of Small Business Management*, 31(4), pp. 60–69.

Carr, D., 1997. "Narrative and the real world: an argument for continuity." In L. Hinchman & S. Hinchman, eds. *Memory, Identity, Community: The Idea of Narrative in the Human Sciences*. New York: State University of New York, pp. 7–25.

Carter, N., Gartner, W. & Reynolds, P., 1996. "Exploring start-up event sequences." *Journal of Business Venturing*, 11, pp. 151–166.

Coleman, J., 1988. "Social capital in the creation of human capital." *American Journal of Sociology*, 94, pp. S95–S120.

Cooper, A., Gimeno-Gascon, J. & Woo, C., 1994. "Initial human and financial capital as predictors of new venture performance." *Journal of Business Venturing*, 9(5), pp. 371–395.

Davidsson, P. & Honig, B., 2003. "The role of social and human capital among nascent entrepreneurs." *Journal of Business Venturing*, 18(3), pp. 301–331.

Davies, T., 2001. "Enhancing competitiveness in the manufacturing sector: key opportunities provided by inter-firm clustering." *Competitiveness Review*, 11(2), pp. 4–15.

Donckels, R. & Lambrecht, J., 1995. "Networks and small business growth: an explanatory model." *Small Business Economics*, 7(4), pp. 273–289.

Doytch, N. & Uctum, M., 2011. "Does the worldwide shift of FDI from manufacturing to services accelerate economic growth? A GMM estimation study." *Journal of International Money and Finance*, 30, pp. 410–427.

Duchesneau, D. & Gartner, W., 1990. "A profile of new venture success and failure in an emerging industry." *Journal of Business Venturing*, 5(5), pp. 297–312.

Elliott, J., 2005. *Using Narrative in Social Research: Qualitative and Quantitative Approaches*, London: Sage.

GEM, 2012. Global Entrepreneurship Monitor Jamaica Report. University of Technology. Kingston: Jamaica.

Granovetter, M., 1983. "The strength of weak ties: a network theory revisited." *Sociological Theory*, 1(1), pp. 201–233.

Hall, G., 1995. *Surviving and Prospering in the Small Firm Sector*, London: Routledge.

Houseman, S., Kurz, C., Lengermann, P. & Mandel, B., 2011. "Offshoring bias in the US manufacturing." *Journal of Economic Perspecitve*, 25(2), pp. 111–132.

Hummels, D., Ishii, J. & Yi, K., 2001. "The nature and growth of vertical specialisation in world trade." *Journal of International Economics*, 54, pp. 75–96.

Jenkins, R. & Sen, K., 2006. "International trade and manufacturing employment in the South: four country case studies." *Oxford Development Studies*, 34(3), pp. 299–322.

Jovanovic, B., 1982. "Selection and the evolution of industry." *Econometrica*, 50, pp. 649–670.

Karim, M., Smith, A. Halgamuge, S. & Islam, M., 2008. "A comparative study of manufacturing practices and performance variables." *International Journal of Production Economics*, 112, pp. 841–859.

Kiss, A., Danis, W. & Cavusgil, S., 2012. "International entrepreneurship research in emerging economies: a critical review and research agenda." *Journal of Business Venturing*, 27(2), pp. 266–290.

Kozicki, S., 1997. "The productivity growth slowdown: diverging trends in the manufacturing and service sectors." Federal Reserve Bank of Kansas City, *Economic Review*, First Quarter, pp. 1–16.

Larsson, E., Hedelin, L. & Garling, T., 2003. "Influence of expert advice on expansion goals of small businesses in rural Sweden." *Journal of Small Business Management*, 41(2), pp. 205–212.

Lee, H., Kelley, D., Lee, J. & Lee, S., 2012. "SME survival: the impact of internationalization, technology, resources and alliances." *Journal of Small Business Management*, 50(1), pp. 1–19.

Lerner, M., Brush, C. & Hisrich, R., 1997. "Israeli women entrepreneurs: an examination of factors affecting performance." *Journal of Business Venturing*, 12(4), pp. 315–339.

Liao, J. & Welsh, H., 2005. "Roles of social capital in venture creation: key dimensions and research implications." *Journal of Small Business Management*, 43(4), pp. 345–362.

Nataraj, S., 2011. "The impact of trade liberalisation on productivity: evidence from India's formal and informal manufacturing sectors." *Journal of International Economics*, 85, pp. 292–301.

Newman, C., Rand, J. & Tarp, F., 2011. "Industry switching in developing countries." *Oxford Journals*, The World Bank Economic Review, pp. 1–32.

Page, J., 2009. "Seizing the day? The global economic crisis and African manufacturing." The Brooking Institutions, pp. 1–23.

Penrose, E., 1959. *The Theory of the Growth of the Firm*, New York: John Wiley.

Pilat, D., Cimper, C., Olsen, K. & Webb, C., 2006. "The changing nature of manufacturing in OECD economies," *STI Working paper 2006/9*.

Ramirez, C., Patel, M. & Blok, K., 2005. "The non-energy intensive manufacturing sector-An energy analysis relating to the Netherlands." *Energy*, 30, pp. 749–767.

Semrau, T. & Werner, A., 2012. "The two sides of the story -network investments and new venture creation." *Journal of Small Business Management*, 50(1), pp. 159–180.

Sharp, J., Irani, Z. & Desai, S., 1999. "Working towards agile manufacturing in the UK industry." *International Journal of Production Economies*, 6, pp. 155–169.

Terziovski, M., 2010. "Innovation practice and its performance implications in small and medium enterprises (SMEs) in the manufacturing sector: a resource-based view." *Strategic Management Journal*, 31, pp. 892–902.

Thomas, L. & D'Aveni, R., 2009. "The changing nature of competition in the U.S. manufacturing sector, 1950–2002." *Strategic Organization*, 7(387), pp. 387–431.

Tregenna, A., 2011. "Manufacturing, productivity, deindustrialization and reindustrialization." Working paper/World Institute for Development Economics Research, pp. 1–25.

Tybout, J., 2000. "Manufacturing firms in developing countries: how well do they do and why?" *Journal of Economic Literature*, 38(1), pp. 11–44.

Watson, J., 2007. "Modelling the relationship between networking and firm performance." *Journal of Business Venturing*, 22, pp. 852–874.

Williams, D.A., 2014. Resources and failure of SMEs: another look. *Journal of Developmental Entrepreneurship*, 19(1), pp. 1–15.

Yin, R., 2003. *Case Study Research: Design and Methods* 3rd ed., London: Sage.

Index

automotive sector, manufacturing, 14–15

Bailey, Albert, 25, 32, 33
Belize, 116
  demographics, 67–8
  Hot Mama's, 67–82
brand building, products, 118–19
Burger King, 22, 23, 24, 34n4
business challenges, Yono Industries, 46–7
business growth
  Caribbean Flavours and Fragrances, 102–3
  forward thinking, 122–3
  Hot Mama's, 70–3
  Island Moldings, 53–6
  Spur Tree Spices, 23–5, 26–7
business strategy
  forward thinking, 122–3
  product knowledge, 119–20
  Spur Tree Spices, 29–30, 32–3
  Yono Industries, 48–9
business survival, network and performance, 6–9

Caribbean Dreams, 86, 92, 93, 117
Caribbean Flavours and Fragrances, 100–101, 113
  business origin and expansion, 102–3
  corporate governance, 108–10
  customers and markets, 105–8

doing business environment, 111–13
employee relations, 108–10
future of business, 110–13
industry conditions, 101–2
networks, 117
products, production and quality control, 103–5
case development, 16–19
  data analyses, 17–18
  data collection, 17
  literature review, 17
  selection criteria, 16
Chanderpaul, Shivnarine, 23
Chicken Supreme, 23, 24, 34n4
company growth, see business growth
corporate governance
  Caribbean Flavours and Fragrances, 108–10
  Hot Mama's, 79–81
  Island Moldings, 60–2
  Perishables Jamaica Ltd., 94–5
  Spur Tree Spices, 30–1
  Yono Industries, 42–3
customers
  Caribbean Flavours and Fragrances, 105–8
  Hot Mama's, 76–8
  Island Moldings, 58–60
  meeting needs of, 120–2
  Perishables Jamaica Ltd., 92–3
  Spur Tree Spices, 25
  Yono Industries, 39, 43–5

doing business, 5, 8, 112, 115
  Caribbean Flavours and Fragrances, 111–13
  Hot Mama's, 82
  Island Moldings, 52, 60, 63
  Perishables Jamaica Ltd., 96–7
  Spur Tree Spices, 23, 31
  Yono Industries, 46–7

economy
  engine of growth, 9–10
  impact of manufacturing sector on, 11–13
  impact of small firms, 2–3
efficiency argument, 115, 124
entrepreneurship research, 4
  firm survival, 8–9
  resources and survival, 4–9
  theory, 1, 19
employee relations
  Caribbean Flavours and Fragrances, 108–10
  Hot Mama's, 79–81
  Island Moldings, 60–2
  Perishables Jamaica Ltd., 94–5
  Spur Tree Spices, 26, 27, 32
  Yono Industries, 45–6
energy costs, manufacturing, 15
engine of growth, 9–10
entrepreneurs
  forward thinking, 122–3
  market diversification, 120–2
  networks, 116–18
entrepreneurship research
  literature, 2, 17
  resources leading to survival or failure, 4–9
  survival and failure, 3–4
expansion, *see* business growth; market expansion
export markets
  Caribbean Flavours and Fragrances, 105–6
  Hot Mama's, 78–9
  Island Moldings, 57, 59–60, 62
  market diversification, 120–2

Perishables Jamaica Ltd., 85, 86, 89, 92–3
Spur Tree Spices, 32
Yono Industries, 44–5

financing
  Island Moldings, 63–4
  Perishables Jamaica Ltd., 96
firm effect, 14
first order narratives, 18
FOOD Ltd. (Fuller Oldham Overseas Development), 69, 72, 83n6, 83n8
fragrance industry, Yono Industries, 36–7

GDP (gross domestic product), 10, 11, 13, 115
Belize, 67–8
Jamaica, 22–3
St Kitts and Nevis, 53
Global Competitiveness Index, 52
Global Competitiveness Report, 23
Golden Krust franchise, 23, 24, 25, 34n4

HACCP (Hazard Analysis Critical Control Procedures), 75–6, 94, 103, 110
Hawkins, Dennis, 23, 24, 33
Hendrickson, Donald, 52, 53–55, 58, 60–2, 64
Hot Mama's, 67, 82
  business environment, 81–2
  business growth and expansion, 72–3
  corporate governance, 79–81
  demographics if Belize, 67–8
  employee relations, 79–81
  export markets, 78–9
  future of business, 81–2
  market expansion and customers, 76–8
  networks, 116, 118
  origins and expansion of business, 68–72
  products, production and quality control, 73–6
Hylton, Honourable Anthony, 31, 34n6

DOI: 10.1057/9781137500328.0013

industry effect, 14
industry switching, 14
international trade, manufacturing sector, 10–11, 13–14
Island Grill, 23, 24, 25, 34n4, 117
Island Moldings, 51, 52, 64
  brand building, 118
  business environment and financing, 63–4
  business origin and expansion, 53–6
  corporate governance, 60–2
  country (St Kitts and Nevis) profile, 52–3
  employee relations, 60–2
  future of business, 62–4
  networks, 117, 118
  product, production and quality control, 56–8
  product knowledge, 120

Jagnarine, Mohan, 23, 24, 27, 33–4
Jamaica
  country information, 22–3
  Perishables Jamaica Ltd., 85–97
  Spur Tree Spices, 22–34
  tea industry in, 85–6
  Yono Industries, 36–50
Jamaica Exporters Association (JEA) Pioneering Industry, 97
James, Anand, 23, 25, 100, 102, 104, 106, 110, 113
Jones, Andre, 36, 37, 42, 43, 45, 46, 47, 50n2

Kaldor's Law, 10
Kennedy, Grace, 27, 78, 106, 117
Kittitian Hill Development, 59, 62, 65n7, 117, 118
Kyoto Protocol, 15

limited portfolio, 19, 115
  entrepreneurship research, 4
  firm survival, 8–9
  market diversification, 121–2
  resources, 4–9

manufacturing sector
  Caribbean region, 115
  case development, 16–19
  changes and developments in, 13–16
  definitional issues, 9–10
  impact on economy, 11–13
  international trade and, 10–11
  overview of, 9–16
  services sector and, 13
  SMEs in, 11
Marie Sharp, 76–8
market expansion
  Hot Mama's, 76–8
  Island Moldings, 58–60
  Spur Tree Spices, 27–30
markets
  Caribbean Flavours and Fragrances, 105–8
  diversification, 120–2
  Island Moldings, 58–60
  Perishables Jamaica Ltd., 92–3
  Yono Industries, 43–5
market structure, 4, 8–9, 19

networking, 7–8, 117
networks
  business survival, 5–6
  lessons learned, 116–18
  performance and, 6–9

OECD countries, manufacturing and economy, 11
offshoring, 14
Oldham, Wilana, 67, 68–9, 82

Perishables Jamaica Ltd., 85, 97–8
  business origin and expansion, 86–90
  corporate governance, 94–5
  customers and market, 92–3
  doing business environment, 96–7
  employee relations, 94–5
  financial management, 96
  future of business, 95–6
  networks, 117

Perishables Jamaica Ltd – *Continued*
   products, production and quality
     control, 90–2
   tea industry in Jamaica, 85–6
production
   Caribbean Flavours and Fragrances,
     103–5
   efficiency, 115, 119, 124
   Hot Mama's, 73–6
   Island Moldings, 56–8
   Perishables Jamaica Ltd., 90–2
   Spur Tree Spices, 26
   Yono Industries, 39, 40–2
products
   brand building, 118–19
   Caribbean Flavours and Fragrances,
     103–5
   Hot Mama's, 73–4
   Island Moldings, 56–8
   knowledge of, 119–20
   Perishables Jamaica Ltd., 90–2
   Spur Tree Spices, 26
   Yono Industries, 39–40

quality control
   Caribbean Flavours and Fragrances,
     103–5
   Hot Mama's, 73–6
   Island Moldings, 56–8
   Perishables Jamaica Ltd., 90–2
   Spur Tree Spices, 26
   Yono Industries, 39–40

research and development
   Perishables Jamaica Limited, 86
   Yono Industries, 38, 45
resource-based argument, firm
   survival or failure, 4–9

services, manufacturing sector, 13
small firms
   entrepreneurship research, 3–4
   impact on national economies,
     2–3
SMEs, manufacturing sector, 11

Spur Tree Spices, 21, 22, 33–4
   business environment, 30–1
   business expansion and growth, 26–7
   business profitability and strategy,
     29–30
   corporate governance, 30–1
   country (Jamaica) information,
     22–3
   factors for survival, 34
   future of business, 31–2
   market expansion, 27–30
   networks, 117, 118
   origins and growth of, 23–5
   product knowledge, 120
   products, production and quality
     control, 26
   strategic planning, 32–3
St Kitts and Nevis
   country profile, 52–3
   Island Moldings, 52–4
strategic planning, Spur Tree Spices,
   29–30, 32–3
sub-contracting, 15
Super Clubs, 40, 50n1
survival/failure, resources leading to,
   4–9

tea industry, Jamaica, 85–6
Tetley Tea, 86–8, 89, 90
T. Geddes Grant, 24, 87
trade barriers, lowering, 14
training
   Caribbean Flavours and Fragrances,
     110, 111
   Hot Mama's, 75
   Island Moldings, 59, 63
   Perishables Jamaica Ltd., 94
   Spur Tree Spices, 26, 32
   Yono Industries, 37, 45–6

vertical specialization, 14

Wright, Norman, 85, 86, 88, 91, 94

Yono Corporation, 37, 39, 43, 50n2

Yono Industries, 35, 36, 49–50
  challenges of doing business, 46–7
  company information, 37–9
  corporate governance, 42–3
  customer and marketing, 43–5
  employee engagement and training, 45–6
  future of business, 48–9
  networks, 118
  production process, 40–2
  product knowledge, 120
  products, production and quality control, 39–40
  sub-sector information, 36–7

The manufacturer's authorised representative in the EU is Springer Nature Customer Service Centre GmbH, Europaplatz 3, 69115 Heidelberg, Germany. If you have any concerns regarding our products, please contact ProductSafety@springernature.com

Printed and bound by CPI Group (UK) Ltd, Croydon, CR0 4YY

23/03/2026

02076355-0018